Environmental Microbiology: Advanced Research and Multidisciplinary Applications

Edited by

Arun Karnwal
Lovely Professional University
India

&

Abdel Rahman Mohammad Said Al-Tawaha
Al Hussein Bin Talal University
Jordan

Environmental Microbiology:
Advanced Research and Multidisciplinary Applications

Editors: Arun Karnwal & Abdel Rahman Mohammad Said Al-Tawaha

ISBN (Online): 978-1-68108-958-4

ISBN (Print): 978-1-68108-959-1

ISBN (Paperback): 978-1-68108-960-7

Published by Bentham Science Publishers Pte. Ltd. Sharjah, UAE. All Rights Reserved.

First published in 2022.

need for a court order if at any point you breach any terms of this License Agreement. In no event will any delay or failure by Bentham Science Publishers in enforcing your compliance with this License Agreement constitute a waiver of any of its rights.

3. You acknowledge that you have read this License Agreement, and agree to be bound by its terms and conditions. To the extent that any other terms and conditions presented on any website of Bentham Science Publishers conflict with, or are inconsistent with, the terms and conditions set out in this License Agreement, you acknowledge that the terms and conditions set out in this License Agreement shall prevail.

Bentham Science Publishers Ltd.
Executive Suite Y - 2
PO Box 7917, Saif Zone
Sharjah, U.A.E.
Email: subscriptions@benthamscience.org

BENTHAM SCIENCE

CONTENTS

PREFACE

Environmental microbiology is a subfield of microbiology that studies the role of microorganisms in maintaining a healthy, viable, and habitable environment. The role of micro-organisms aims to foster a prosperous, viable and inhabited environment. Microbes are considered to have both negative and positive effects on the environment, as their pollution can cause severe health problems, while on the other hand, they have various beneficial uses such as organic material degradation, being a source of nutrients in food chains, nutrient recycling, and pollutant bioremediation. This book offers an introduction to the discipline of environmental microbiology. It also demonstrates the importance of environmental microbes in our daily lives. It describes the main microorganisms that are found in environmental microbiology, their methodological options, and possible human impact. There is more to the successful exploitation of a given environment than can be explained exclusively in terms of environmental microbiology. An important thrust in this book is the new challenges of modern environmental microbiology, where pathogens and bioremediation are still important topics. This book is the result of hard work and the many efforts of the different authors. We pay them a warm thanks. This book provides a comprehensive review on environmental microbiology. The chapters, each written by subject matter specialists, help scientists, teachers, students, extension workers, farmers, consumers, administrators, traders and NGOs in increasing their understanding of environmental microbiology. This book on environmental microbiology comprises 10 chapters. The topics of our chapters are as follows: (1) Environmental microbiology: introduction and scope, (2) Impact of microbial diversity on the environment, (3) Rhizospheric microbial communication, (4) Microbial communication: a significant approach to understand microbial activities, and interactions, (5) Nutrient cycling: an approach to environmental sustainability, (6) Microbial biosensors for environmental monitoring, (7) Microbial degradation, bioremediation and biotransformation, (8) Bioremediation of hazardous organics in industrial refuses, (9) Role of microbial biofilms in bioremediation, (10) Microbial processing for the valorization of waste and application. Internationally, the interdisciplinary and multifactor global modern system of teamwork has been recognized for scientific excellence. Therefore, most chapters have involved the collaboration of 4–8 or more diverse international authors from different countries. Thus the book represents a truly global perspective consistent with the nature of environmental microbiology.

Arun Karnwal
Lovely Professional University
India
&

Abdel Rahman Tawaha
Al Hussein Bin Talal University
Jordan

List of Contributors

Abdel Rahman M. Tawaha	Department of Biological Sciences, Al Hussein bin Talal University, Maan, Jordan
Abdel Razzaq Al-Tawaha	Department of Crop Science, Faculty of Agriculture, Universiti Putra Malaysia, 43400 Serdang, Selangor, Malaysia
Abdul Basit	Department of Plant Pathology, Agriculture College, Guizhou University, Guiyang 550025, P.R. China
Abdur Rauf	Department of Chemistry, University of Swabi, Anbar, Khyber Pakhtunkhwa, Pakistan
Abhijit Dey	Department of Life Sciences, Presidency University, Kolkata, India
Arun Karnwal	Department of Microbiology, School of Bioengineering and BioSciences, Lovely Professional University, Phagwara, India
Amanullah	Department of Agronomy, the University of Agriculture, Peshawar, Pakistan
Amrita Kumari Rana	Department of Microbiology, College of Basic Sciences and Humanities, Punjab Agricultural University, Ludhiana-141004, Punjab, India
Aftab Ahmad	Institute of Home & Food Sciences, Government College University, Faisalabad, Pakistan
Dhriti Kapoor	Department of Botany, School of Bioengineering and Biosciences Phagwara, Lovely Professional University, Punjab, India
Dhriti Sharma	Department of Botany, School of Bioengineering and Biosciences Phagwara, Lovely Professional University, Punjab, India Sidharth Govt. College, Nadaun, Hamirpur, Himachal Pradesh, India
Farhan Saeed	Institute of Home & Food Sciences, Government College University, Faisalabad, Pakistan
Hiba Alatrash	Department of Field Crops, General Commission for Scientific Agricultural Research, Aleppo, Syria
Ifrah Usman	Institute of Home & Food Sciences, Government College University, Faisalabad, Pakistan
Imran	Department of Agronomy, the University of Agriculture, Peshawar, Pakistan
Kunwarpreet Kaur	Department of Microbiology, College of Basic Sciences and Humanities, Punjab Agricultural University, Ludhiana-141004, Punjab, India
Khursheed Hussain	Division of Vegetable Science, SKUAST-Kashmir, India
Muhammad Afzaal	Institute of Home & Food Sciences, Government College University, Faisalabad, Pakistan
Manpreet Kaur Somal	Department of Biotechnology, School of Bioengineering and Biosciences, Lovely Professional University, Phagwara-144411, Punjab, India
Mukesh Kumar	Department of Microbiology, School of Bioengineering and Biosciences, Lovely Professional University, Phagwara-144411, Punjab, India
Muhammad Nouman	University Institute of Food Sciences & Technology, University of Lahore, Lahore, Pakistan

Mamta Pujari	Department of Botany, School of Bioengineering and Biosciences Phagwara, Lovely Professional University, Punjab, India
Muhammad Saeed	National Institute of Food Science & Technology, University of Agriculture, Faisalabad, Pakistan
Muhammad Manhal Awad Al-Zoubi	General Commission for Scientific Agricultural Research, Administration of Natural Resource Research, Syria
Nafiaah Naqash	Department of Zoology, School of Bio-engineering and Bio-sciences, Lovely Professional University, Phagwara-144411, Punjab, India
Navneet	Department of Botany and Microbiology, Gurukul Kangri University, Haridwar–249404, Uttar-akhand, India
Neelesh Babu	Department of Microbiology, BFIT Group of Institutions, Chakrata Road, Suddhowala, Dehra-dun–248007, Uttarakhand, India
Nujoud Alimad	Damascus University, Faculty of Agriculture, Damascus, Syria
Palani Saranraj	Department of Microbiology, , Sacred Heart College (Autonomous), Tirupattur 635601, Tamil Nadu, India
Pratibha Vyas	Department of Microbiology, College of Basic Sciences and Humanities, Punjab Agricultural University, Ludhiana-141004, Punjab, India
Rahul Singh	Department of Zoology, School of Bio-engineering and Bio-sciences, Lovely Professional University, Phagwara-144411, Punjab, India
Ravindra Soni	Department of Agricultural Microbiology, College of Agriculture, Indira Gandhi Krishi Vishwa-vidyalaya, Raipur–492012, India
Renu Bhardwaj	Department of Botanical and Environmental Sciences, Guru Nanak Dev University, Amritsar, Punjab, India
Ritu Bala	Department of Microbiology, School of Bioengineering and Biosciences, Lovely Professional University, Phagwara-144411, Punjab, India
Riham Fouzi Zahalan	General Commission for Scientific Agricultural Research, Administration of Natural Resource Research, Syria
Rohan Samir Kumar Sachan	Department of Microbiology, School of Bioengineering and Biosciences, Lovely Professional University, Phagwara-144411, Punjab, India
Sameena Lone	Division of Vegetable Science, SKUAST-Kashmir, India
Samia Khanum	Department of Botany, University of the Punjab, Lahore, Pakistan
Savita Bhardwaj	Department of Botany, School of Bioengineering and Biosciences Phagwara, Lovely Professional University, Punjab, India
Shah Khalid	Department of Agronomy, the University of Agriculture, Peshawar, Pakistan
Shiv Gautam	Serve India Inter College, Roshanpur, Gadarpur Udham Singh Nagar–262401, Uttarakhand, India
Sufiara Yousuf	Department of Zoology, School of Bio-engineering and Bio-sciences, Lovely Professional University, Phagwara-144411, Punjab, India

CHAPTER 1

Environmental Microbiology: Introduction and Scope

Dhriti Sharma[1,2,#]**, Savita Bhardwaj**[1,#]**, Mamta Pujari**[1]**, Renu Bhardwaj**[3] **and Dhriti Kapoor**[1,*]

[1] *Department of Botany, School of Bioengineering and Biosciences, Lovely Professional University, Phagwara, Punjab, India*

[2] *Sidharth Govt. College, Nadaun, Hamirpur, Himachal Pradesh, India*

[3] *Department of Botanical and Environmental Sciences, Guru Nanak Dev University, Amritsar, Punjab, India*

Abstract: Environmental microbiology deals with the role of microorganisms in supporting a thriving, viable and inhabitable environment. It helps to figure out the nature and functioning of the microbial population residing in all parts of the biosphere, *i.e.*, air, water, and soil. Microbes are known to affect the environment both negatively and positively, as their contamination may lead to serious health issues on one hand, whereas various welfare activities like degradation of organic material, being a source of nutrients in food chains, recycling of nutrients, and bioremediation of pollutants are also associated with them on the other hand. In a way, their practical importance makes them a special tool in the hands of environment microbiologists to lessen the deleterious impact of different environmental problems. The degradation potential of microbes earns them a place in treating wastewater, containing organic and inorganic impurities being originated in public and industrial arenas whereby minerals, nutrients, and a number of other eco-friendly by-products are also generated. Microbial species like *Pseudomonas*, *Sphingomonas*, and *Wolinella* are few among those species which are commonly engaged in this process of degradation of harmful effluents being continuously added into the environment, thus ensuring the safety and sustenance of the latter.

Furthermore, their degradative abilities also help them to effectively confront and conquer the problem of oil spillage in sea waters resulting in less ecological damage. The manipulation of microbes in the present times has gained quite an important place in our lives in which this discipline of environmental microbiology contributes by unraveling all such possibilities of utilizing the microbes to our benefit. The present chapter provides a deep insight into this important branch of microbiology and its scope, which will help better understand its role in other fields such as agriculture, medicine, pharmacy, clinical research, and chemical and water industries.

[*] **Corresponding author Dhriti Kapoor:** Department of Botany, School of Bioengineering and Biosciences, Lovely Professional University, Phagwara, Punjab, India; E-mails: dhriti405@gmail.com
[#] Contributed equally.

Arun Karnwal & Abdel Rahman Mohammad Said Al-Tawaha (Eds.)

Keywords: Begradation, Biotic Interactions, Bioremediation, Environmental Microbiology, Human Welfare, Nutrient Cycling, Wastewater Treatment.

INTRODUCTION

The invisible world of microorganisms, belonging to three principal life realms-Archaea, Bacteria, and Eukaryota and viruses- has played a pivotal role in the evolutionary process of the rest of the organisms dwelling on earth [1]. Being the earliest life forms, they have brought about major changes in the primitive reducing atmosphere; turning it into an oxidizing one with the help of oxygenic photosynthesis. Further, by developing adaptive mechanisms, they have colonized almost all the inhabitable areas on the earth, even those that offer the most unusual and extreme circumstances in terms of temperature, pressure, salinity, radiation, and pH [2]. The intriguing cosmopolitan nature, diversity, and immensity coupled with longevity have made the microbes' interactions with their surroundings more interesting. The study of these interactions between microorganisms and macroorganisms, including their environment, has now been upgraded to a new discipline of 'Environmental Microbiology' or 'Applied Microbial Ecology'.

Environmental impacts of microbial activities are beneficial as well as harmful. A plethora of essential ecosystem services are carried out by these microbes; all the existing life forms and the biosphere at large, exhibit direct or indirect dependency upon the microbial activities. The microorganisms regulate biogeochemical cycles around the globe *via* having a major say over the important assimilative processes like fixation of Carbon and Nitrogen along with metabolism of Sulphur, Methane, *etc.* [3]. On the other hand, the baneful aspects of microbial existence in our surroundings involve the decomposition of our food items, textiles, and dwellings; disease in animals and crop plants. The credit for most of the progress made so far in this field goes to the advent of new technologies like the availability of radioisotopes, chemical-sensitive microelectrodes, and cultivation-independent techniques. New branches like metagenomics and metatranscriptomics plus metaproteomics wherein sequencing of the complete DNA complement recovered from the environment and quantification of actual expression of genetic potential are performed, have taken the science of environmental microbiology to a massive leap forward [4].

History of Environmental Microbiology

Microbe-based studies date back as early as the seventeenth century when an amateur, Anton von Leeuwenhoek reported their existence and named them animalcules. However, initially, research activities regarding microbes were carried out only in the context of a physiological perspective without giving much importance to the ecological aspects involved. This is revealed in the works of Louis Pasteur and M. Beijerinck who studied the distribution of microbes and invented the enrichment culture technique for microbes [5]. Among a few others, S. Winogradsky attempted microbial studies keeping their medical aspects aside; developed the Winogradsky column, and discovered chemosynthesis. He is credited to be among the first students of Environmental Microbiology [6]. But, Hungate and co-workers in the 20th century pioneered this new discipline of environmental microbiology by including quantitative aspects of ecological activities performed by microbes. In its initial stages, the center point of environmental microbiology was on public health, owing to a number of microbial disease outbreaks caused due to contaminated food and water like food poisoning, typhoid, cholera, *etc*. However, in the 1960s, a famous literary work 'Silent Spring' by Rachel Carlson shifted this focus to the presence of chemical pollutants in natural resources and their ill effects on health. This eventually led to the discovery of clean-up mechanisms by employing microbes and a whole new aspect of environmental microbiology *i.e.*, bioremediation. Further, the inclusion of molecular genetics and the advent of other biotechnological applications have modernized this field.

Biotic Interactions- The Basis of the Science of Environmental Microbiology

Microbial diversity inhabiting almost all the existent habitats on the earth is a cumulative result of key interactions exhibited by microorganisms among themselves and with macroorganisms. These interactive associations hint at the co-evolution of the partners involved making them well-adapted and specialized (Table **1**). Major biotic interactions of microbes are summarized as follows:

1) Symbiosis: It is a type of biotic interaction where microbes, particularly bacteria, get involved with other microbes or organisms of higher groups. Though microbes are quite small, but they contribute significantly to the physiological and evolutionary processes of eukaryotes [7]. Further, the symbiotic relationships exhibited by microbes can be categorized into a) Mutualism b) Commensalism c) Ammensalism, which in turn cast major impacts on the ecosystem of which they are a part [8].

a) Mutualism: It comprises the mutually beneficial relationship between the involved microbes and their partners [9]. Besides lichens (alga + fungus) and mycorrhiza (fungus + roots of plants), a classic example of microbial mutualism is a consortium formed between a methane-producing archaebacteria (*Methanobacterium omelianskii)* and an ethyl alcohol fermenting organism where the latter provides hydrogen to the former so that proper growth and production of methane occurs, this process is known as cross-feeding or Syntrophy [10].

Ethanol fermenting partners exhibit thermodynamically unfavoured endergonic reactions but the association with archaebacterial partners turns the nature of the overall reaction into an exergonic one, thus their existence in extreme environments is ensured which is not possible for them to do individually [11, 12]. In ecosystems with inadequate energy and nutrient resources like deep subsurfaces of soil or water bodies, this type of mutualistic interaction is believed to assist the growth and survival of microorganisms along with the production of energy in higher quantities [13, 14].

b) Commensalism: This symbiotic relation is quite common where one microbial partner benefits from the other partner's metabolic products without the latter exhibiting any good or bad impact. For example, two microbial species- fungus *Saccharomyces cerevisiae* (A fungus) support the growth of bacteria *Proteus vulgaris* in a mixed culture by providing it with niacin like growth factor, which is not possible in growing the bacteria in pure culture; the fungal partner is neither harmed nor benefitted from this interaction [15]. Some of the microbial flora thriving upon different body parts of human beings also come under the helm of commensalistic interactions like *E. coli* residing in the intestine.

c) Ammensalism: It is a type of antagonistic interaction wherein one partner is negatively affected while the other one exhibits indifference to this relationship by remaining unaffected. A peculiar example substantiating this interaction is between the microbial organisms *Lactobacillus casei* and *Pseudomonas taetrolens* where the former owing to the by-products synthesized during the production of lactic acid inhibits the growth of the latter *via* inducing a reduction in the overall yield of its main product *i.e.* lactobionic acid without itself getting least affected in terms of growth and behavior [16]. One more such example is between *Staphylococcus xylosus* and *Kocuria varians* involved in the fermentation of meat and vegetables and the former casts an inhibitory effect on the latter [17].

2) Parasitism: In negative interaction, the smaller partner, called the parasite, derives nutrition and shelter from, in, or, on the body of the larger partner, called the host, and casts an inhibitory effect on the survival of the host. A parasite is unable to exist without its host, however, the parasitism can be temporary or permanent, external or internal in nature. A large number of microbes find their place in medical microbiology pertaining to their parasitic nature and therefore being the causative agent of several diseases of viral, bacterial, fungal, or protozoan origin. In terms of microbial interactions, the viruses- bacteriophages parasitize upon bacteria, especially those involved in the fermentation of food items [18].

Table 1. Biotic interactions among microbes and their role in the environment.

Name of Interaction	Nature of Interaction	Examples	Ecological Importance
Positive Interactions			
Mutualism	Beneficial for both the partner species involved	Lichens (an alga and a fungus) are pioneers of xeric succession in the ecosystem	These are the pioneers of xeric succession in the ecosystem
Proto-cooperation	Beneficial for both the partner species but not obligatory	Association of *Desulfovibrio vulgaris* and *Chromatium vinosum*	A non-obligatory partnership between the carbon cycle and Sulphur cycle
Commensalism	One of the partner species is benefitted without affecting the other one	*Flavobacterium* (host) and *Legionella pneumophila* (commensal)-the latter derives cysteine from the host to survive in aquatic habitat.	It derives cysteine from the host to survive in aquatic habitat
Syntrophism	Beneficial for both the partner species, here growth of one organism either depends or improved by the other one which itself also benefits from the association, not possible when alone	Methanogenic ecosystem in a sludge digester between *Mehanobacterium* and other fermentative bacteria.	Plays a vital role in anaerobic environments and converts complex organic molecules into CH_4 and CO_2
Negative Interactions			
Ammensalism	One of the partner species is harmed by the other which itself remains indifferent to the association	*Thiobacillus thiooxidans* this Sulphuric acid producing bacteria lowers pH and inhibits growth of other bacteria in culture media	Such unilateral interactions have been proven to increase community stability
Parasitism	One Partner species (host) is negatively affected by the other one (parasite) which is usually smaller in size	*Bdellovibrio* is parasitic to many gram –ve bacteria	Parasites shape community structure *via* affecting trophic interactions, competition, biodiversity and keystone species
Predation	One Partner species (prey) is negatively affected by the other one (predator) which is usually larger in size	Many protozoans present in the soil feeding on bacterial populations; thus maintaining the count of soil bacteria at optimum level	Predation increases the biodiversity in ecosystem by preventing dominance of single species
Competition	Both the partner species compete for a common resource	*Paramecium caudatum* and *P. aurelia*, where the latter outcompetes the former one in terms of growth	Plays an important role in ecology and evolution of better species

Importance of Microorganisms

Owing to their cosmopolitan nature, a plethora of roles are performed by microorganisms in the environment. All interactions occur between these microbes and macroorganisms (from symbiotic, neutral, commensalistic, exploitative, and competitive) which are discussed under two broad heads i) Useful activities and ii) Harmful activities.

Useful Activities

Nutrient Cycling

An intricate complex of microorganisms (bacteria, fungi, protists) is the potential source of nutrients to their biotic surroundings (macro-organisms like plants and animals) instead of soil, as considered in plant physiology [19, 20]. The microbial richness in the environment has been found to promote plant growth *via* different mechanisms like modulating the hormone signaling inside the plants, outsmarting or expelling the disease-producing microbes, and adding nutrients into the soil [21, 22].

In the last activity, nutrients like Carbon, Nitrogen, Phosphorus and sulfur, *etc.* which are otherwise organically held in living forms get released and added back into the ecosystem upon their death and decay through the process of biodegradation by the microorganisms (mostly *via* saprotrophic bacteria and fungi).They are then converted into more favored ionic forms such as ammonium, nitrate, phosphate, and sulfate as depicted in Fig. (**1**) [23]. All the major biogeochemical cycles (Carbon cycle, Nitrogen cycle, Sulphur, and Phosphorus cycle) rely on this biodegrading potential of microbes for their completion [24] For example, the nitrifying and ammonifying bacteria (*Nitrosomonas, Nitrosococcus, Nitrocystis,* and *Bacillus* sp.) release and perform the fixation of the naturally unavailable form of molecular nitrogen into the preferred and bioavailable nutrient forms. This nutrient cycling followed by transformation boosts the growth of plants and in turn those dependent upon plants. The overall productivity of an ecosystem, in a way, depends upon the microbial activities in the recycling of nutrients [25].

Chemosynthesis

Microbes are crucial to all the world's ecosystems for ensuring their sustenance and survivability, especially the ones where the possibility of photosynthesis is ruled out owing to the absence of light such as in deep marine biomes and hydrothermal vents. In these areas, microbes such as *Beggiatoa, Ferrobacillus, Gallionella,* and *Thiovirga* provide nutrition to other organisms by performing chemosynthesis. The genus *Sulfurihydrogenibium* has maximum carbon dioxide fixation rates in the dark at high temperatures whereas *Thiovirga sufurooxydans* bacterium excels at comparatively lower temperatures and shows chemolithotrophic activity to produce energy by oxidizing sulphur, sulfide, and thiosulphate [26]. Moreover, some of these chemotrophic microbes can also perform well under anoxygenic environments.

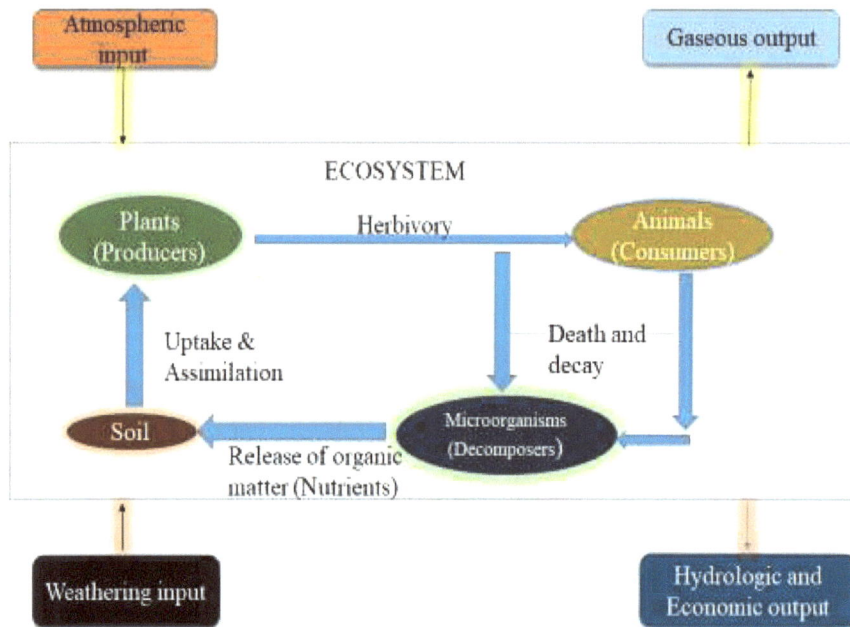

Fig. (1). A generalized scheme of nutrient cycling in an ecosystem depicting the important role of microbes.

Soil Formation

Microorganisms residing in the soil serve in multiple ways. The process of formation of soil, however, involves the contribution of biological, physical, and chemical factors but microbes perform the major role. The microbes are the chief driving force behind many transformations in the reservoir pool of nutrients of the soil and they also give rise to stable and labile forms of carbon, affect the formation of bedrocks, and enhance the soil porosity and glomalin content which in turn help in the establishment of subsequent plant communities [27].

Contribution to the Evolutionary Process

Major transfer of genes in a horizontal manner occurs in microbial populations which holds evolutionary significance [28, 29]. Evidence of this horizontal gene transfer came forth first from Griffith's transformation experiment in 1928 in which the *Pneumococcus* bacteria got modified from the non-virulent form into a virulent one upon horizontally taking up the genes from its close relatives. Other methods of recombination in prokaryotes like conjugation and transduction also work upon the principle of horizontal gene transfer. Moreover, during the evolution of prokaryotic and the eukaryotic domain, extensive horizontal transfer has taken place, which will help restructure phylogenetic trees while comparing these with the ones constructed based on genome histories created from the fossil records [30, 31].

Harmful Activities

Spoilage of Food Products

Every year huge economic losses are incurred due to food spoilage with the significant contribution of microorganisms [32]. As per the reports of USDA Economic Research Service estimates, food close to 96 billion pounds of weight in the United States alone, is rendered unfit for human consumption either at retail, food service, or consumer level of marketing. The Flavour and shape of these consumables are changed due to microbial activity and about 25% of global food production is estimated to be lost due to this spoilage [33]. Further, the demand for fresh and pesticide-free food items, along with enhanced shelf life, has left the food articles more prone to microbial. The rotting of vegetables, fruits, meat, bread, or souring of milk and milk products is caused by saprotrophic bacteria, which are always present in the air and settle down on exposed food articles. Given their extremely small size, food infestation with microbes, especially bacteria and yeast, is hard to notice except for molds. The nature of food also contributes to this spoilage by microorganisms like the food items with more water content (meat, milk, and seafood) are spoiled comparatively more often as compared to those with less water content. However, the cereals constituting our staple diet are spoiled mainly by fungi like molds and yeasts. Common examples of food spoiling bacteria and fungi are *Acinetobacter, Brochothrix, Clostridium, Flavobacterium, Micrococcus, Pseudomonas, Staphylococcus*, lactic acid bacteria, members of *Enterobacteriaceae, Aspergillus flavus, Aspergillus niger, Penicillium and Rhizopus* sp [34].

Spoilage of Household Products

Domestic articles like textiles, paper, plastic, paint optical instruments, leather, canvas, and wooden articles are also exposed to microbial spoilage. Bacteria like *Spirochaete cytophaga, Cellulomonas* sp., and fungi like *Alternaria, Aspergillus, Chaetomium, Cladosporium, Penicillium, and Rhizopus* contribute majorly to the deterioration of these articles.

Diseases in Plants and Animals

Microbes are also pathogenic, causing serious inflictions to the biotic component of the ecosystem. The majority of the plant and animal diseases are of bacterial origin in about 90% of human diseases are caused by bacteria. Some of the examples of bacterial, fungal, and viral diseases are summarised in Table **2**.

Table 2. **Major pathogens of microbial origin cause diseases in plants, animals, and human beings.**

Pathogen Host	Bacteria	Fungi	Virus
Plants	Citrus canker (*Xanthomonas citri*); Crown gall of Apple (*Agrobacterium tumefaciens*); Soft rot of potato (*Erwinia carotovora*)	Apple scab (*Venturia inaequalis*); Black stem rust of wheat (*Puccinia graminis tritici*); Late blight of potato (*Phytophthora infestans*)	Tobacco mosaic (Tobacco mosaic virus); Yellow vein mosaic of Okra (Hibiscus virus I); Leaf curl of tomato (Leaf curl tomato virus)
Animals	Anthrax (*Bacillus anthracis)* Black leg of cattle (*Clostridium chauvei*); Tularemia (*Francisella tularensis)*	Blastomycosis (*Blastomyces* sp.); Candidiasis (*Candida albicans*); Mucormycosis (*Mucor, Rhizopus*)	Foot and mouth disease (Foot and mouth virus); Rinderpest (Rinderpest virus); Rabies (Rabies virus)
Human beings	Cholera (*Vibrio cholerae*); Tyhoid (*Salmonella typhi*); Tuberoculosis (*Mycobacterium tuberoculosis*); Plague (*Yersinia pestis*)	Dermatomycosis (*Microsporum* spp., *Trichophyton* spp.); Apergillosis (*Aspergillus fumigatus*)	Influenza Influenza virus), Herpes (Herpes virus), AIDS (Human immunodefeciency virus) Measles (Measles virus), Polio (Polio virus), Rabies (Rabies virus), Small pox (Small pox virus)

Scope of Environmental Microbiology

The emerging field of environmental microbiology has a plentitude of scope which in turn makes its applications extensive in diverse fields like industrial microbiology, soil microbiology, food safety, diagnostic microbiology, aquatic microbiology, water industries, safely disposing off hazardous wastes, biotechnology, occupational health/infection control, and aero microbiology. Here, a few of them are described to give a clear idea about the important role the microbes play in preserving the environment and human welfare at large.

Bioremediation: Besides being quite efficient in degrading naturally occurring material substances, the microorganisms have also been found to decompose some chemically synthesized compounds known as xenobiotics. Increased cognizance regarding the harmful impacts of these chemical pollutants produced as by-products in food, agricultural, chemical and pharmaceutical industries and are continuously added to the environment has kindled the research activities centered on manufacturing easily degradable substances or the techniques which help in degrading the contaminants in an eco-friendly manner. The application of diverse microbes individually or in a collective manner for this purpose by utilizing their biodegradation potential is known as bioremediation. Both *in situ* (on the actual site) and *ex-situ* (away from the actual site of contamination) strategies are practiced in it. Metal biosorption, biostimulation, bioaugmentation, and bioventing are some of the *in situ* techniques, whereas landfarming, biopiling, and composting come under the helm of *ex-situ* remediation. Still, some of these methods can be adopted both *in situ* and *ex-situ* conditions, so cannot be demarcated into one type.

Out of all of these, biostimulation stands apart in terms of its wide application and advantages; in it, the growth of otherwise naturally occurring microorganisms is boosted through the external supplementation with nutrients to help them to degrade pollutants more effectively [35]. Biostimulation can be concertedly used with a related technique of bioaugmentation in which the microbes with high degradative potential are inoculated into the affected site to fasten the process of remediation. The biotic interactions between the microbes and the environment affect the degradative abilities of remedial techniques which work together on the substrate (contaminant).

Examples of some of the microorganisms possessing biodegradation potential for contaminants are:

i. Arhaebacteria like *Halobacterium, Haloferax, Halococcus*;
ii. Bacteria like *Pseudomonas putida, Dechloromonas aromatica, Nitrosomonas europaea, Nitrobacter hamburgensis, Deinococcus radiodurans, Sphingomonas, Wolinella*;
iii. Fungi like White rot fungi- *Phanerochaete chrysosporium, Pleurotus ostreatus* and *Trametes versicolor* [36 - 38].

Some Pollutant-specific Bioremediation Techniques

Degradation of Oil Spills

All the major economies of the world witness competition for the most valuable energy source of present times *i.e.* petroleum oil [39]. However, during the multiple stages of petroleum oil production, refining, processing, and at the time of its storage and transportation, there is an imminent danger of oil spill accidents which result in environmental degradation [40, 41]. Oil spills in peculiar environments such as deep-sea areas, deserts, polar regions, and wetlands, further aggravate the difficulty level of this problem. But this knotty issue of oil contamination in the ecosystem can be readily rectified by the use of petroleum hydrocarbon-degrading bacteria like *Achromobacter, Alkanindiges, Dietzia, Enterobacter, Mycobacterium, Pandoraea* [42 - 45]. Types, requirements in different environments, and the basic process of bioremediation of oil spills are briefly summarised in Fig. (**2**).

Biomineralization

New mineral technologies involve the use of microorganisms for on-site sequestering of inorganic pollutants such as those present in acid mine drainages [46, 47]. Microbes affect the process of mineral formation in various ways like they can either coprecipitate or simply adsorb the inorganic metals. For example, the Iron oxidizing bacteria change the ferrous (Fe^{2+}) form of iron into ferric (Fe^{3+}), which precipitates more easily, forming ferrihydrites for trapping inorganic pollutants [48, 49].

Several different bacterial strains have been recognized of Fe-oxidizing bacteria (*Gallionella ferruginea, Acidithiobacillus ferrooxidans*) and Fe- and As-oxidizing bacteria (*Thiomonas* sp.) which assist in Iron and Arsenic holding mineral phases for biomineralization. Some of the other microbes have been reported to synthesize Manganese oxides to sequester these pollutants [50, 51].

Microscopic studies have revealed the direct connection between microbial cells and mineral phases where the trapping of inorganic pollutants leading to biomineralization occurs either inside the periplasm or at the extracellular level *via* the formation of extracellular polymeric substances [52]. Microbial phosphatases are also employed to trap metal pollutants like Cr, U, Pb, and Sr through oxidizing organic phosphates which release harmless inorganic phosphates by precipitation [53 - 55]. Though these practices are quite economical and promising, bioremediation still needs to be popularised on a larger scale.

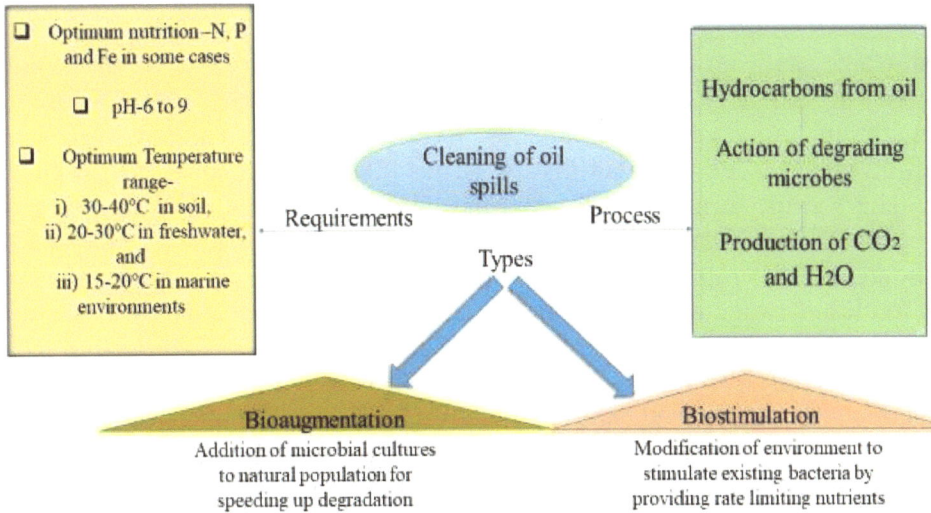

Fig. (2). Summing up different aspects of cleaning of oil spills by microbes.

Biofertilizers

Rhizosphere colonizing microorganisms, especially bacteria and fungi which enhance the availability of primary nutrients to the target plant crops are categorized as biofertilizers. These are also known as plant growth-promoting microbes which can be of several types like Nitrogen fixers, Phosphate solubilizers, Potash, and Zinc mobilizers, depending upon the activity they perform inside the soil. Live formulations or carrier-based inoculums of a number of microbes when applied to different plant parts like roots, and seeds, or simply added to the soil increase the phytoavailability of soil nutrients by changing them from unstable to accessible forms. Some major examples of biofertilizers are photosynthetic bacteria, nitrogen-fixing bacteria plus cyanobacteria (*Rhizobium, Nostoc, Anabaena*), and other bacteria like *Actinomyces, Azotobacter, Bacillus, Pseudomonas, Lactobacillus* sp. along with microbes of fungal origin such as yeast, *Trichoderma* and mycorrhizal associations. *Rhizobium* is one of the extensively studied nitrogen-fixing bacteria in legume plants whereas *Azospirillum* fixes the nitrogen in cereal and fodder crops besides enhancing the water and mineral uptake rate. These biofertilizers improve the crop yield sustainably besides being cost-effective, handy to use, harmless to the environment, renewable, and eco-friendly [56]. All these features make them an irreplaceable and essential part of the Integrated Nutrition System (INS) and Integrated Plant Nutrition System (IPNS) [57].

Wastewater Treatment

Wastewater has been generated in huge quantity primarily due to anthropogenic actions, for instance, increased industrial effluents, agronomic techniques, and urbanization; however, simultaneous release of wastewater without proper handling creates severe contamination issues [58]. Bioaerosol is diffused into the atmosphere in the form of small water particles, which act as microbe transporters during wastewater remediation [59]. Various microbes are recognized as potential candidates for the remediation of wastewater, *e.g. Lipomyces starkeyi* yeast was used to clean sewage disposal and produce lipids which are then used as a moiety for the generation of biodiesel [60]. Yeasts such as *Rhodotorula glutinis*, *Rhorosporidium toruloides*, and *Cryptococcus curvatus* can also be used to remediate wastewater [61]. The combination of different microbes, such as yeast and algae, is also an important technique for biological wastewater treatment [62]. This treatment is of two major types- i) Aerobic and ii) Anaerobic depending upon the requirement of oxygen, a comparative account of these two is summarised in Table **3**. Anaerobic remediation methods have also been used for the salinity removal from wastewater *e.g. Haloferax denitrificans*, and *H. mediterranei* removed nitrate and nitrite from salinity water sources [63]. The co-immobilized application of *Chlorella vulgaris* and *Pseudomonas putida* exhibited efficient exclusion of ammonium, phosphate, and organic carbon, showing that the nutrient accumulation capacity of *C. vulgaris* and *P. putida* was increased from wastewater when used in combination [64]. Microbial immobilized media was produced by combining polyvinyl alcohol, polyethylene glycol, and activated sludge which showed 99% nitrification and denitrification activity and eliminated 85% of the organic matter from wastewater [65]. *Bacillus salmalaya* 139SI bacteria improved the removal efficiency of Cr^{6+} and increased crude oil - waste's water solubility [66, 67].

Table 3. Showing major differences between two types of biological wastewater treatment.

Aerobic Waste Water Treatment	Parameter	Anaerobic Waste Water Treatment
-Microbial activity takes place in the presence of oxygen -Degradation end products are CO_2, H_2O and biomass	Bases of the Process	-Microbial activity takes place in the presence of oxygen -Degradation end products are CO_2, CH_4 and biomass
Suitable for wastewater that is difficult to biodegrade like municipal and refinery sewage with COD less than 1000ppm *i.e.* the one having low to medium organic impurities	Applicability	Suitable for wastewater which is easily biodegradable like food and beverage industry sewage rich in sugar/starch/alcohol with COD more than 1000ppm *i.e.* the one having medium to high organic impurities
Relatively fast	Reaction kinetics	Relatively slow

(Table 3) cont.....

Aerobic Waste Water Treatment	Parameter	Anaerobic Waste Water Treatment
Comparatively more	Net sludge production	Comparatively less
Not much needed, can be directly discharged or simply filtered/disinfected	After-treatment	Needs to be followed by aerobic treatment
Relatively low	Cost effectiveness	Relatively more economical
i)Suspended Film Processes *e.g.* Activated Sludge like Membrane bioreactor (MBR), Extended Aeration Oxidation Ditch (EAOD); ii)Fixed Film Processes like biotower/trickling filter, Biological Aerated Filter (BAF), Moving Bed Biofilm Bioreactor (MBBR) or iii)Hybrid processes *e.g.* Integrated Fixed Film activated sludge (IFAS)	Example Technologies	Digesters / Continuously Stirred Tank reactors (CSTR), Upflow Anaerobic Sludge Blanket (USAB), Fluidized Bed Reactors (FBR), Ultra High Rate Fluidized Bed Reactors *e.g.* ICTM

Biofilms

Since the beginning of microbiology, microbes have been mainly recognized as planktonic, free-floating cells and designated based on their growth, phenological, and biochemical aspects in nutritionally rich culture media. However, in all major natural ecosystems, these microorganisms are usually seen in tight interaction with substrates in the form of multicellular masses, known as biofilms, which are attached to the mucus they release [68]. During biofilm formation, microbe assembly is immobilized on a solid structure *via* electrostatic, hydrophobic, and covalent bonds carried out by the microbe itself and its constituents, for instance, cilia, fimbriae, cell wall, and EPS [69]. Biofilms are found almost in all humid ecosystems where surface attachment source is available and proper nutrient amount is accessible. A biofilm comprises an individual bacterial moiety as well as also made up of several bacteria, fungi, algae, and protozoa species, in which around 97% of the biofilm matrix is either H_2O or solvent, the physical characteristics of which are controlled by the solutes added in it [70]. In a biofilm, mass transference relies upon diffusion whereas, its thickness relies upon the potential of constituents and O_2 penetration [70].

Bioenergy Production

The energy generated from biofuels and waste contributes to 10% of the global energy demand compared to other resources. Hence, biofuels are recognized as vital sources for future sustainable energy generation, and bioenergy generation *via* microbes is considered a promising source. For example, microbes like microalgae, bacteria, fungi, and yeast have the potential to produce CH_4, H_2 gas, biodiesel, or bioethanol [71] (Fig. **3**). *Nannochloropsis gaditana* produced biodiesel through transesterification with methanol under supercritical circumstances with the highest amount of 0.48 g g^{-1} biodiesel at a temperature of 255–265 °C and a reaction period of 50 min [72]. Microalgae biomass of *Chlorella* sp. was pre-treated with anaerobic acterium *Bacillus licheniformis* for the production of biogas, where methane generation was enhanced from 9.2 to 22.7% by altering the amount of *B. licheniformis* [73].

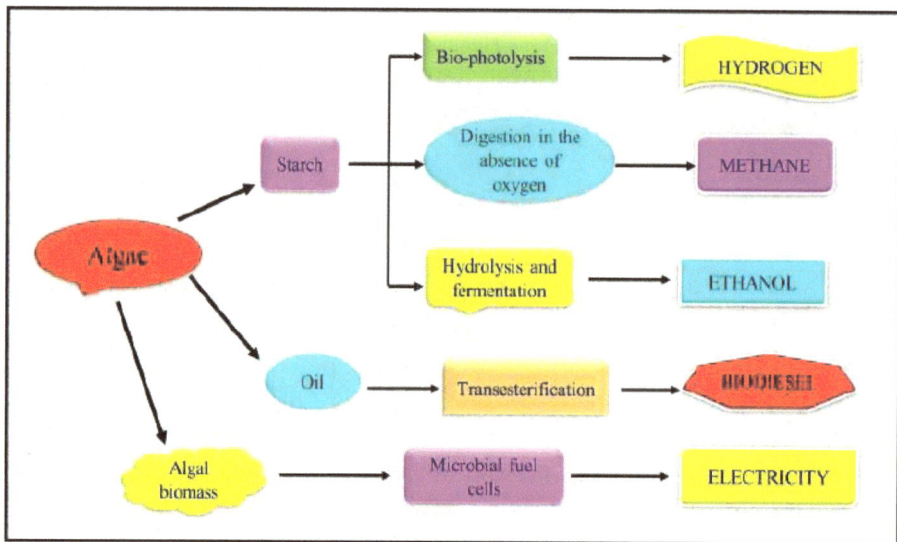

Fig. (3). Bioenergy production by using microbes (algae) (modified from Baicha *et al.* [71]).

Oleaginous microbes, for instance, microalgae, yeasts, and bacteria are also utilized as an alternative technique for biodiesel generation due to their characteristics like decreased land demand, small farming time, and also accumulation of greater amounts of lipids in their cells through utilizing organic carbon sources *i.e.* carbohydrates and organic acids [74]. *Clostridium butyricum* uses POME as a substrate produced H_2 at a pH of 5.5 with an H_2 amount of 3.2 L per liter of POME [75]. Anaerobic bacteria *Clostridium paraputrificum* significantly converted *N*-acetyl-*D*-glucosamine to H_2 with an amount of hydrogen 2.2 mol H_2/mol of GlcNAc [76]. Certain bacterial and fungal strains improve the formation of biogas by triggering the action of specific enzymes, for example, cellulolytic strains like *Actinomycetes* enhance biogas formation in the range of 8.4–44% [77].

CONCLUSION

Though invisible, the microorganisms are all around us, residing in every possible habitat, and have largely contributed to making the environment what it is today by helping in the evolutionary process of the rest of the organisms and transforming the primitive anoxygenic conditions on this planet. These earliest life forms continue to rule over us by causing serious disease outbreaks, spoiling a large number of economically important articles but at the same time also act as saviors by giving us provisions essential for sustaining life and preserving the environment. This particular arena dealing with applied part of microorganisms in our surroundings has helped in its elevation to a new discipline of environmental microbiology. Biotic and abiotic components of the ecosystem are in a way regulated by the activities of these microorganisms. Microbes interact among themselves and with other microorganisms in various manners and confer the environment with their beneficial and baneful services. Anthropogenic interventions have interfered with the ecosystem's natural harmony, giving rise to the problem of pollution in almost all the natural resources found in soil, air, and water. Over the years, environmental microbiology has become a science that shares its interfaces with a large number of specialties like agriculture, industries, and health and offers services that can rectify this imbalance in the environment. Bioremediation, biofertilizers, bioenergy, biofilms, and biological sewage treatment are some of the emerging and unique services which use microorganisms to preserve the environment and assist its other biotic inhabitants to lead a healthy and quality life.

CONSENT FOR PUBLICATION

Not applicable.

CONFLICT OF INTEREST

The author declares no conflict of interest, financial or otherwise.

ACKNOWLEDGEMENT

Declared none.

REFERENCES

[1]   Barton LL, Northup DE. Microbial Ecology. Wiley-Blackwell. Oxford: John Wiley & Sons., New York 2011; 22.
        [http://dx.doi.org/10.1002/9781118015841]

[2]   Bowler C, Karl DM, Colwell RR. Microbial oceanography in a sea of opportunity. Nature 2009; 459(7244): 180-4.
        [http://dx.doi.org/10.1038/nature08056] [PMID: 19444203]

[3]   Lupp C. Microbial oceanography. Nature 2009; 459(7244): 179.
         [http://dx.doi.org/10.1038/459179a] [PMID: 19444202]

[4]   Shakya M, Lo CC, Chain PSG. Advances and challenges in metatranscriptomic analysis. Front Genet
         2019; 10: 904.
         [http://dx.doi.org/10.3389/fgene.2019.00904] [PMID: 31608125]

[5]   Konopka A. What is microbial community ecology? ISME J 2009; 3(11): 1223-30.
         [http://dx.doi.org/10.1038/ismej.2009.88] [PMID: 19657372]

[6]   Madigan M, Brock T. Biology of microorganisms. 13th ed., San Francisco: Benjamin Cummings
         2012.

[7]   López-García P, Eme L, Moreira D. Symbiosis in eukaryotic evolution. J Theor Biol 2017; 434: 20-33.
         [http://dx.doi.org/10.1016/j.jtbi.2017.02.031] [PMID: 28254477]

[8]   Krasner RI. The Microbial Challenge: Science, Disease, and Public Heatlh. Jones & Bartlett
         Publishers 2009.

[9]   Srivastava S. Understanding bacteria. Springer Science & Business Media 2003.
         [http://dx.doi.org/10.1007/978-94-017-0129-7]

[10]   Faust K, Raes J. Microbial interactions: from networks to models. Nat Rev Microbiol 2012; 10(8):
         538-50.
         [http://dx.doi.org/10.1038/nrmicro2832] [PMID: 22796884]

[11]   Kirchman DL. Processes in microbial ecology. Oxford University Press 2018.
         [http://dx.doi.org/10.1093/oso/9780198789406.001.0001]

[12]   Bronstein JL. The study of mutualism.Mutualism. USA: Oxford University Press 2015.
         [http://dx.doi.org/10.1093/acprof:oso/9780199675654.003.0001]

[13]   Schink B, Stams AJ. Syntrophism among prokaryotes. The prokaryotes 2006.
         [http://dx.doi.org/10.1007/0-387-30742-7_11]

[14]   Lau MCY, Kieft TL, Kuloyo O, *et al.* An oligotrophic deep-subsurface community dependent on
         syntrophy is dominated by sulfur-driven autotrophic denitrifiers. Proc Natl Acad Sci USA 2016;
         113(49): E7927-36.
         [http://dx.doi.org/10.1073/pnas.1612244113] [PMID: 27872277]

[15]   Veiga JP. Commensalism. Amensalism, and Synnecrosis. Encyclopedia of Evolutionary Biology
         2016; pp. 322-8.

[16]   García C, Rendueles M, Díaz M. Synbiotic fermentation for the co production of lactic and
         lactobionic acids from residual dairy whey. Biotechnol Prog 2017; 33(5): 1250-6.
         [http://dx.doi.org/10.1002/btpr.2507] [PMID: 28556559]

[17]   Tremonte P, Succi M, Reale A, Di Renzo T, Sorrentino E, Coppola R. Interactions between strains of
         Staphylococcus xylosus and Kocuria varians isolated from fermented meats. J Appl Microbiol 2007;
         103(3): 743-51.
         [http://dx.doi.org/10.1111/j.1365-2672.2007.03315.x] [PMID: 17714408]

[18]   Sturino JM, Klaenhammer TR. Engineered bacteriophage-defence systems in bioprocessing. Nat Rev
         Microbiol 2006; 4(5): 395-404.
         [http://dx.doi.org/10.1038/nrmicro1393] [PMID: 16715051]

[19]   Bonkowski M, Villenave C, Griffiths B. Rhizosphere fauna: the functional and structural diversity of
         intimate interactions of soil fauna with plant roots. Plant Soil 2009; 321(1-2): 213-33.
         [http://dx.doi.org/10.1007/s11104-009-0013-2]

[20]   Müller DB, Vogel C, Bai Y, Vorholt JA. The plant microbiota: systems-level insights and
         perspectives. Annu Rev Genet 2016; 50(1): 211-34.
         [http://dx.doi.org/10.1146/annurev-genet-120215-034952] [PMID: 27648643]

[21]   Mendes R, Garbeva P, Raaijmakers JM. The rhizosphere microbiome: significance of plant beneficial, plant pathogenic, and human pathogenic microorganisms. FEMS Microbiol Rev 2013; 37(5): 634-63.
[http://dx.doi.org/10.1111/1574-6976.12028] [PMID: 23790204]

[22]   Verbon EH, Liberman LM. Beneficial microbes affect endogenous mechanisms controlling root development. Trends Plant Sci 2016; 21(3): 218-29.
[http://dx.doi.org/10.1016/j.tplants.2016.01.013] [PMID: 26875056]

[23]   van der Heijden MGA, Bardgett RD, van Straalen NM. The unseen majority: soil microbes as drivers of plant diversity and productivity in terrestrial ecosystems. Ecol Lett 2008; 11(3): 296-310.
[http://dx.doi.org/10.1111/j.1461-0248.2007.01139.x] [PMID: 18047587]

[24]   Fenchel T, Blackburn H, King GM, Blackburn TH. Bacterial biogeochemistry: the ecophysiology of mineral cycling. 2012.

[25]   Schimel JP, Bennett J. Nitrogen mineralization: challenges of a changing paradigm. Ecology 2004; 85(3): 591-602.
[http://dx.doi.org/10.1890/03-8002]

[26]   Yang T, Lyons S, Aguilar C, Cuhel R, Teske A. Microbial communities and chemosynthesis in yellowstone lake sublacustrine hydrothermal vent waters. Front Microbiol 2011; 2: 130.
[http://dx.doi.org/10.3389/fmicb.2011.00130] [PMID: 21716640]

[27]   Schulz S, Brankatschk R, Dümig A, Kögel-Knabner I, Schloter M, Zeyer J. The role of microorganisms at different stages of ecosystem development for soil formation. Biogeosciences 2013; 10(6): 3983-96.
[http://dx.doi.org/10.5194/bg-10-3983-2013]

[28]   Smets BF, Barkay T. Horizontal gene transfer: perspectives at a crossroads of scientific disciplines. Nat Rev Microbiol 2005; 3(9): 675-8.
[http://dx.doi.org/10.1038/nrmicro1253] [PMID: 16145755]

[29]   McDaniel LD, Young E, Delaney J, Ruhnau F, Ritchie KB, Paul JH. High frequency of horizontal gene transfer in the oceans. Science 2010; 330(6000): 50.
[http://dx.doi.org/10.1126/science.1192243] [PMID: 20929803]

[30]   Keeling PJ, Palmer JD. Horizontal gene transfer in eukaryotic evolution. Nat Rev Genet 2008; 9(8): 605-18.
[http://dx.doi.org/10.1038/nrg2386] [PMID: 18591983]

[31]   Andersson JO. Horizontal gene transfer between microbial eukaryotes. Horizontal Gene Transfer 2009; pp. 473-87.
[http://dx.doi.org/10.1007/978-1-60327-853-9_27]

[32]   Dousset X, Jaffrès E, Zagorec M. Spoilage: Bacterial Spoilage. Encyclopedia of Food and Health 2016; 106-12.

[33]   Bondi M, Messi P, Halami PM, Papadopoulou C, de Niederhausern S. Emerging microbial concerns in food safety and new control measures. BioMed Res Int 2014; 2014: 1-3.
[http://dx.doi.org/10.1155/2014/251512] [PMID: 25110665]

[34]   Doulgeraki AI, Ercolini D, Villani F, Nychas GJE. Spoilage microbiota associated to the storage of raw meat in different conditions. Int J Food Microbiol 2012; 157(2): 130-41.
[http://dx.doi.org/10.1016/j.ijfoodmicro.2012.05.020] [PMID: 22682877]

[35]   Mrozik A, Piotrowska-Seget Z. Bioaugmentation as a strategy for cleaning up of soils contaminated with aromatic compounds. Microbiol Res 2010; 165(5): 363-75.
[http://dx.doi.org/10.1016/j.micres.2009.08.001] [PMID: 19735995]

[36]   Brim H, McFarlan SC, Fredrickson JK, *et al.* Engineering Deinococcus radiodurans for metal remediation in radioactive mixed waste environments. Nat Biotechnol 2000; 18(1): 85-90.
[http://dx.doi.org/10.1038/71986] [PMID: 10625398]

[37] Fragoeiro S. Use of fungi in bioremediation of pesticides Applied Mycology Group Institute of Bioscience and Technology. Cranfield University 2005.

[38] Singh H. Mycoremediation: fungal bioremediation. John Wiley & Sons 2006; pp. 283-5.
[http://dx.doi.org/10.1002/0470050594.ch8]

[39] Yuanyuan S. On-site management of international petroleum cooperation projects. Natural Gas Exploration and Development 2009; p. 02.

[40] Chen M, Xu P, Zeng G, Yang C, Huang D, Zhang J. Bioremediation of soils contaminated with polycyclic aromatic hydrocarbons, petroleum, pesticides, chlorophenols and heavy metals by composting: Applications, microbes and future research needs. Biotechnol Adv 2015; 33(6): 745-55.
[http://dx.doi.org/10.1016/j.biotechadv.2015.05.003] [PMID: 26008965]

[41] Wang C, Liu X, Guo J, Lv Y, Li Y. Biodegradation of marine oil spill residues using aboriginal bacterial consortium based on Penglai 19-3 oil spill accident, China. Ecotoxicol Environ Saf 2018; 159: 20-7.
[http://dx.doi.org/10.1016/j.ecoenv.2018.04.059] [PMID: 29730405]

[42] Margesin R, Labbé D, Schinner F, Greer CW, Whyte LG. Characterization of hydrocarbon-degrading microbial populations in contaminated and pristine Alpine soils. Appl Environ Microbiol 2003; 69(6): 3085-92.
[http://dx.doi.org/10.1128/AEM.69.6.3085-3092.2003] [PMID: 12788702]

[43] Lea-Smith DJ, Biller SJ, Davey MP, *et al.* Contribution of cyanobacterial alkane production to the ocean hydrocarbon cycle. Proc Natl Acad Sci USA 2015; 112(44): 13591-6.
[http://dx.doi.org/10.1073/pnas.1507274112] [PMID: 26438854]

[44] Varjani SJ. Microbial degradation of petroleum hydrocarbons. Bioresour Technol 2017; 223: 277-86.
[http://dx.doi.org/10.1016/j.biortech.2016.10.037] [PMID: 27789112]

[45] Xu X, Zhai Z, Li H, Wang Q, Han X, Yu H. Synergetic effect of bio-photocatalytic hybrid system: g-C3N4 and Acinetobacter sp. JLS1 for enhanced degradation of C16 alkane. Chem Eng J 2017; 323: 520-9.
[http://dx.doi.org/10.1016/j.cej.2017.04.138]

[46] Barkay T, Schaefer J. Metal and radionuclide bioremediation: issues, considerations and potentials. Curr Opin Microbiol 2001; 4(3): 318-23.
[http://dx.doi.org/10.1016/S1369-5274(00)00210-1] [PMID: 11378486]

[47] Adriano DC, Wenzel WW, Vangronsveld J, Bolan NS. Role of assisted natural remediation in environmental cleanup. Geoderma 2004; 122(2-4): 121-42.
[http://dx.doi.org/10.1016/j.geoderma.2004.01.003]

[48] De Yoreo JJ, Vekilov PG. Principles of crystal nucleation and growth. Rev Mineral Geochem 2003; 54(1): 57-93.
[http://dx.doi.org/10.2113/0540057]

[49] Hohmann C, Winkler E, Morin G, Kappler A. Anaerobic Fe(II)-oxidizing bacteria show as resistance and immobilize as during Fe(III) mineral precipitation. Environ Sci Technol 2010; 44(1): 94-101.
[http://dx.doi.org/10.1021/es900708s] [PMID: 20039738]

[50] Bargar JR, Fuller CC, Marcus MA, *et al.* Structural characterization of terrestrial microbial Mn oxides from Pinal Creek, AZ. Geochim Cosmochim Acta 2009; 73(4): 889-910.
[http://dx.doi.org/10.1016/j.gca.2008.10.036]

[51] Spiro TG, Bargar JR, Sposito G, Tebo BM. Bacteriogenic manganese oxides. Acc Chem Res 2010; 43(1): 2-9.
[http://dx.doi.org/10.1021/ar800232a] [PMID: 19778036]

[52] Benzerara K, Morin G, Yoon TH, *et al.* Nanoscale study of As biomineralization in an acid mine drainage system. Geochim Cosmochim Acta 2008; 72(16): 3949-63.

[http://dx.doi.org/10.1016/j.gca.2008.05.046]

[53]   Templeton AS, Trainor TP, Spormann AM, *et al.* Sorption *versus* biomineralization of Pb(II) within Burkholderia cepacia biofilms. Environ Sci Technol 2003; 37(2): 300-7.
[http://dx.doi.org/10.1021/es025972g] [PMID: 12564901]

[54]   Goulhen F, Gloter A, Guyot F, Bruschi M. Cr(VI) detoxification by Desulfovibrio vulgaris strain Hildenborough: microbe–metal interactions studies. Appl Microbiol Biotechnol 2006; 71(6): 892-7.
[http://dx.doi.org/10.1007/s00253-005-0211-7] [PMID: 16896506]

[55]   Beazley MJ, Martinez RJ, Sobecky PA, Webb SM, Taillefert M. Uranium biomineralization as a result of bacterial phosphatase activity: insights from bacterial isolates from a contaminated subsurface. Environ Sci Technol 2007; 41(16): 5701-7.
[http://dx.doi.org/10.1021/es070567g] [PMID: 17874776]

[56]   Mishra P, Dash D. Rejuvenation of biofertilizer for sustainable agriculture and economic development. Consilience 2014; (11): 41-61.

[57]   Raghuwanshi R. Opportunities and challenges to sustainable agriculture in India. NeBio 2012; 3(2): 78-86.

[58]   Gonçalves AL, Pires JCM, Simões M. A review on the use of microalgal consortia for wastewater treatment. Algal Res 2017; 24: 403-15.
[http://dx.doi.org/10.1016/j.algal.2016.11.008]

[59]   Cyprowski M, Stobnicka-Kupiec A, Ławniczek-Wałczyk A, Bakal-Kijek A, Gołofit-Szymczak M, Górny RL. Anaerobic bacteria in wastewater treatment plant. Int Arch Occup Environ Health 2018; 91(5): 571-9.
[http://dx.doi.org/10.1007/s00420-018-1307-6] [PMID: 29594341]

[60]   Angerbauer C, Siebenhofer M, Mittelbach M, Guebitz GM. Conversion of sewage sludge into lipids by Lipomyces starkeyi for biodiesel production. Bioresour Technol 2008; 99(8): 3051-6.
[http://dx.doi.org/10.1016/j.biortech.2007.06.045] [PMID: 17719773]

[61]   Pirozzi D, Ausiello A, Zuccaro G, Sannino F, Yousuf A. Culture of oleaginous yeasts in dairy industry wastewaters to obtain lipids suitable for the production of II-generation Biodiesel. World Acad Sci Eng Technol 2014; 7(4): 26.

[62]   Chi Z, Zheng Y, Jiang A, Chen S. Lipid production by culturing oleaginous yeast and algae with food waste and municipal wastewater in an integrated process. Appl Biochem Biotechnol 2011; 165(2): 442-53.
[http://dx.doi.org/10.1007/s12010-011-9263-6] [PMID: 21567213]

[63]   Oren A. Industrial and environmental applications of halophilic microorganisms. Environ Technol 2010; 31(8-9): 825-34.
[http://dx.doi.org/10.1080/09593330903370026] [PMID: 20662374]

[64]   Shen Y, Gao J, Li L. Municipal wastewater treatment *via* co-immobilized microalgal-bacterial symbiosis: Microorganism growth and nutrients removal. Bioresour Technol 2017; 243: 905-13.
[http://dx.doi.org/10.1016/j.biortech.2017.07.041] [PMID: 28738545]

[65]   Shin D, Yoon S, Park C. Biological characteristics of microorganisms immobilization media for nitrogen removal. J Water Process Eng 2019; 32: 100979.
[http://dx.doi.org/10.1016/j.jwpe.2019.100979]

[66]   Dadrasnia A, Chuan Wei K, Shahsavari N, Azirun M, Ismail S. Biosorption potential of *Bacillus salmalaya* strain 139SI for removal of Cr (VI) from aqueous solution. Int J Environ Res Public Health 2015; 12(12): 15321-38.
[http://dx.doi.org/10.3390/ijerph121214985] [PMID: 26633454]

[67]   Ismail S, Dadrasnia A. Biotechnological potential of *Bacillus salmalaya* 139SI: a novel strain for remediating water polluted with crude oil waste. PLoS One 2015; 10(4): e0120931.
[http://dx.doi.org/10.1371/journal.pone.0120931] [PMID: 25875763]

[68]   Wimpenny J, Manz W, Szewzyk U. Heterogeneity in biofilms: Table 1. FEMS Microbiol Rev 2000; 24(5): 661-71.
[http://dx.doi.org/10.1111/j.1574-6976.2000.tb00565.x] [PMID: 11077157]

[69]   Dadrasnia A, Usman MM, Lim KT, *et al.* Microbial Aspects in Wastewater Treatment–A Technical. Environmental Pollution and Protection 2017; 2(2): 75-84.

[70]   Sutherland I. The biofilm matrix – an immobilized but dynamic microbial environment. Trends Microbiol 2001; 9(5): 222-7.
[http://dx.doi.org/10.1016/S0966-842X(01)02012-1] [PMID: 11336839]

[71]   Baicha Z, Salar-García MJ, Ortiz-Martínez VM, *et al.* A critical review on microalgae as an alternative source for bioenergy production: A promising low cost substrate for microbial fuel cells. Fuel Process Technol 2016; 154: 104-16.
[http://dx.doi.org/10.1016/j.fuproc.2016.08.017]

[72]   Jazzar S, Olivares-Carrillo P, Pérez de los Ríos A, *et al.* Direct supercritical methanolysis of wet and dry unwashed marine microalgae (Nannochloropsis gaditana) to biodiesel. Appl Energy 2015; 148: 210-9.
[http://dx.doi.org/10.1016/j.apenergy.2015.03.069]

[73]   He S, Fan X, Katukuri NR, Yuan X, Wang F, Guo RB. Enhanced methane production from microalgal biomass by anaerobic bio-pretreatment. Bioresour Technol 2016; 204: 145-51.
[http://dx.doi.org/10.1016/j.biortech.2015.12.073] [PMID: 26773949]

[74]   Cho HU, Park JM. Biodiesel production by various oleaginous microorganisms from organic wastes. Bioresour Technol 2018; 256: 502-8.
[http://dx.doi.org/10.1016/j.biortech.2018.02.010] [PMID: 29478783]

[75]   Chong M, Rahim R, Shirai Y, Hassan M. Biohydrogen production by *Clostridium butyricum* EB6 from palm oil mill effluent. Int J Hydrogen Energy 2009; 34(2): 764-71.
[http://dx.doi.org/10.1016/j.ijhydene.2008.10.095]

[76]   Evvyernie D, Yamazaki S, Morimoto K, *et al.* Identification and characterization of *Clostridium paraputrificum* M-21, a chitinolytic, mesophilic and hydrogen-producing bacterium. J Biosci Bioeng 2000; 89(6): 596-601.
[http://dx.doi.org/10.1016/S1389-1723(00)80063-8] [PMID: 16232804]

[77]   Merlin Christy P, Gopinath LR, Divya D. A review on anaerobic decomposition and enhancement of biogas production through enzymes and microorganisms. Renew Sustain Energy Rev 2014; 34: 167-73.
[http://dx.doi.org/10.1016/j.rser.2014.03.010]

CHAPTER 2

Impact of Microbial Diversity on the Environment

Hiba Alatrash[1,*], Abdel Rahman M. Tawaha[2], Abdel Razzaq Al-Tawaha[3], Samia Khanum[4], Abdur Rauf[5], Arun Karnwal[6], Abhijit Dey[7], Sameena Lone[8], Khursheed Hussain[8], Palani Saranraj[9], Imran[10], Amanullah[10] and Shah Khalid[10]

[1] *Department of Field Crops, General Commission for Scientific Agricultural Research, Aleppo, Syria*

[2] *Department of Biological Sciences, Al Hussein bin Talal University, P.O. Box 20, Maan, Jordan*

[3] *Department of Crop Science, Faculty of Agriculture, Universiti Putra Malaysia, 43400 Serdang, Selangor, Malaysia*

[4] *Department of Botany, University of the Punjab, Lahore, Pakistan*

[5] *Department of Chemistry, University of Swabi, Anbar, Khyber Pakhtunkhwa, Pakistan*

[6] *Department of Microbiology, School of Bioengineering and BioSciences, Lovely Professional University, Phagwara, India*

[7] *Department of Life Sciences, Presidency University, Kolkata, India*

[8] *Division of Vegetable Science, SKUAST-Kashmir, India*

[9] *Department of Microbiology, Sacred Heart College (Autonomous), Tirupattur 635601, Tamil Nadu, India*

[10] *Department of Agronomy, the University of Agriculture, Peshawar, Pakistan*

Abstract: Microbial diversity is an essential aspect of any ecosystem on earth. Microorganisms are the most common and diversified population in the soil. A microbe is a microscopic organism that can be studied in a single-cell or colony. On the other hand, microbes have a positive or negative effect on their surroundings. Microbial diversity plays an essential role in bioremediation, which is the method of detoxifying or neutralizing radioactive waste into less harmful or non-toxic compounds by secreting various bacterial and fungal enzymes. In this chapter, we focus on (i) the impact of microbial diversity on detoxifying pollutants (bioremediation), (ii) microbial role in biofuel production, (iii) microbial role in ore leaching (bioleaching), (iv) microbial role in controlling biogeochemical cycles (v) microbial role in soil quality and agriculture improvement (vi).

* **Corresponding author Hiba Alatrash:** Department of Field Crops, General Commission for Scientific Agricultural Research, Aleppo, Syria; E-mail: hiba.fad.16@gmail.com
\# Contributed equally.

Arun Karnwal & Abdel Rahman Mohammad Said Al-Tawaha (Eds.)

Keywords: Biofuel, Biogeochemical, Bioleaching, Climate Change, Microbial Diversity.

INTRODUCTION

A microbe or microorganism is a microscopic organism found in a single-celled form or a colony of cells. Microbes include two groups, the first group consists of viruses, bacteria, and archaea called prokaryotes. The other group consists of algae, fungi, protists, and protozoa called eukaryotes. Microbes can live in every kind of habitat (terrestrial, aquatic, atmospheric, or living host), and their diversity makes them tolerant to many conditions, such as heat, cold, drought, salinity, and flooding conditions [1 - 3]. Microbial diversity plays a key role in different ecosystems on earth [4].

Microbes are the first active cellular life forms on earth, more than 3.0 billion years before the evolution of multicellular life forms [5]. Robert Hooke made the first discovery of microbes' existence during the period (1635-1703), and Antoni van Leeuwenhoek (1632-1723) described microbes as a kind of life that cannot be seen with the naked eye. In 1729–1799, Spallanzani and Pasteur (1822–1895) disproved the spontaneous microbial generation theory. Followed by the classification of bacteria into four groups based on shape by the German biologist Ferdinand Julius Cohn (1828 –1898), this classification is still in use these days, while Robert Koch's (1843–1910) studies show that microbes have a prominent role in causing illnesses [6, 7].

Microbes can positively or negatively affect their environment, and most people think of microbes as harmful because of their roles in causing infectious diseases to humans, animals, and plants. However, many microbes are beneficial because of their environmental metabolic activities. For example, detoxifying pollutants (bioremediation) by the bacterial enzymes of some microbes, such as *Bacillus*, *Alcanivorax borkumensis*, and *Pseudomonas*, can reduce environmental pollutants [8]. Several microbes can convert animal wastes, biomass, and grain into biofuels such as ethanol and methane [9]. Ore leaching (bioleaching) is the process of extracting metals from ores by converting metals into a soluble form and keeping ores nearly insoluble through bio-oxidation and complexation processes carried out by a group of microorganisms [10] such as *Actinobacteria, Acidithiobacillus, Acidiphilium, Sulfolobus* and *Acidianus* [11]. It controls the biogeochemical cycle of the essential elements, mainly carbon (C), nitrogen (N), and oxygen (O), which build up all living systems [1]. Several novel microbes contribute to the soil quality and crop improvement by playing an essential role in soil–the plant relationship [12] in many ways, such as providing nutrients to plants, *i.e.*, fixing atmospheric nitrogen and soil phosphates, converting insoluble elements into soluble forms so plants can obtain them easily (potassium, phosphates, nitrogen, carbon, and iron), and improving the plant's defense system against environmental stresses [13, 14]. So without microbes, there would be no life on earth [1, 4, 9].

Detoxifying Pollutants (Bioremediation)

Bioremediation is the process of detoxifying or neutralizing hazardous waste into less toxic or non-toxic substances by secreting various bacterial and fungal enzymes [1, 15]. Bioremediation is the most promising, efficient, low-cost, and eco-friendly method to clean up pollution [1, 16, 17]. The prime bioremediation is known as bacteria, archaea, and fungi.

The first commercial use of this process was in Ambler, Pennsylvania (1972), to clean up the spill of the Sun Oil pipeline [18]. Many researches showed that different species of fungi and bacteria such as *Microbacterium, Pseudomonas, Bacillus, Mycobacterium, Phanerochaete, Panibacillus, Arthrobacter, Acinetobacteria, and Sphingomonas* are capable of biodegrading or detoxifying substances through several ways by eating them or by various beneficial enzymes [1, 4] such as Oxidoreductases, Peroxidases, Laccases, Lipases, Protease and Cellulases [19]. These enzymes can clean up environmental pollutants without producing harmful substances [8, 20].

Pseudomonas is a well-studied bacteria that can degrade alkanes, monoaromatics, and naphthalene under aerobic conditions; it is also used in the degradation of oil. *Sphingomonas* and *Mycobacterium* are also used for oil degradation [1]. *Arthrobacter nicotinovorans* is a bacterium capable of degrading a biodegradable pesticide like atrazine and its derivatives like propazine simazine and cyanazine [21]. *Phanerochaete chrysosporium* is a fungus that can clean up different types of pollutants such as lignin, pentachlorophenol, and dioxin [22]. The process of cleaning up and eliminating environmental hazards pollutants such as toxic petroleum and chlorinated compounds by fungi is called mycoremediation [2, 23].

Microbes also can remove toxic heavy metals by reduction *via* different cellular enzymes such as mercury oxidase reduces Hg^{+2} to Hg; consequently, it spreads outside the cell due to its low evaporation point [24]. Microbial cell metabolism plays a key role in heavy metals bioremediation strategy. Microbial growth and development require heavy metals as essential micronutrients in amounts that vary with different types of microbes. The peculiarity of heavy metals is that a low heavy metal concentration promotes microbial growth, while a higher concentration harms the integrity of the cell membrane, cell organelles, and their genetic material [25]. Some microbes transform toxic metal species into less toxic forms through oxidative or reductive metabolism [1].

Bacteria are the most effective microbes used in the remediation of heavy metal-polluted [26]. Several studies reported that various strains of bacteria such as (*Bacillus anthracis, Micrococcus roseus, Klebsiella pneumonia, Lysinibacillus fusiformis, Pseudomonas sp. W6,* and *Vibrio fluvialis*) could eliminate environmental hazards like toxic heavy metals such as arsenic, cadmium, lead and Hg [27 - 32]. Pandey and Bhatt [33] pointed out that the bacterial strain *Exiguobacterium* sp. can remove 99% of arsenic toxicity.

Role of Microbes in Biofuel Production

The dramatic increase in fossil fuel consumption worldwide, the increase in fossil fuel price, and significant concerns about climate change have motivated the search for renewable resources of fuel. Hence, biofuel such as (bioethanol, biodiesel, biobutanol, propanol, and synthetic hydrocarbons) [34, 35] is considered the main alternate for fossil fuels [36 - 38]. The first use of living organisms to obtain biofuel was in 1970s [4].

Biofuel production through the use of microbes has two different processes in terms of synthetic and biological systems: first, converting biomass into sugars for fuel by synthetic microbial enzymes; second, producing fuel directly by novel microbes. Microbes such as (*Bacillus subtilis, Clostridium thermocellum, Corynebacterium glutamicum, Escherichia coli, Enterobacter lignolyticus, Pseudomonas aeruginosa, Rhodotorula glutinis, Saccharomyces cerevisiae; Zymomonas mobilis*) play an essential role in biofuel production [39 - 44].

Escherichia coli bacterium can produce ethanol under anaerobic conditions by an endogenous process. *E. coli* metabolize one mole of glucose into two moles of formate, two moles of acetate, and one mole of ethanol. Also, *S. cerevisiae* and *Zymomonas mobilis* can produce ethanol through fermentation [39, 45]. Other microbes such as *Corynebacterium glutamicum* and *Closteridia* species are capable of producing ethanol depending on the biofuel type and the nature of the targeted material [42, 46].

Microbes' Role in Ore Leaching (Bioleaching)

It is an efficient technology that uses microorganisms such as *Thiobacillus, Leptospirillum,* and *Thermophilic* bacteria in extracting metals from ores. Also, it has the ability to clean up toxic industrial products, sewage sludge, and soil polluted with heavy metals [47, 48]. Bioleaching occurs naturally wherever the conditions are appropriate for microbial growth [49].

Generally, bioleaching is the process of extracting metals from ores by converting metals into a soluble form and keeping ores nearly insoluble using bio-oxidation and complexation processes carried out by a group of microorganisms [10]. Since the 1970s, the number of reported genera and species of microorganisms involved in metal bioleaching has increased substantially [48]. These microorganisms include *Acidithiobacillus, Acidiphilium, Alicyclobacillus, Ferrimicrobium, Pseudomonas, Leptospiriuulm, Sulfobacillus,* and *Thermosulfidooxidans* [11, 50 - 53].

Bioleaching has two mechanisms: direct "contact" and indirect "non-contact" leaching [54, 55]. In direct (contact leaching), the bacterial cells are physically in contact with the metal's surface. The electrochemical processes to convert metals into a soluble form occur at the interface between the metal's surface and the bacterial cells. While the in indirect mechanism (Non-contact leaching), there is no physical contact between the bacterial cells and the metal's surface. Bacteria only have a catalytic function because they accelerate the reoxidation of the metal, which becomes very slow in the absence of the bacteria [47, 56, 57]. The leaching efficiency in both mechanisms mainly depends on the microorganisms' efficiency and the ore's mineralogical and chemical composition to be leached [47].

Microbes Controlling Biogeochemical Cycles

Microbes are the most important pillars in every ecological system on earth by controlling the biogeochemical cycle of essential elements, such as carbon (C), nitrogen (N) and oxygen (O).

In Carbon Cycle (C)

Carbon is the main component of all living organisms. Microbes play an essential role in the carbon cycle; they extract carbon from non-living sources and provide it for living organisms and themselves. The most common form of carbon that enters the carbon cycle is carbon dioxide (CO_2), a water-soluble gas in the atmosphere. Microbes play an essential role in the carbon cycle. Microbes and algae-like plants use CO_2 from the air and the water around them to synthesize sugar to use it as food or store it as insoluble starch. This process is called photosynthesis [1]. Also, heterotrophic bacteria and fungi recycle the nonliving organic material. Moreover, *Actinobacteria* and *Proteobacteria* are involved in the degradation of soluble organic compounds, such as sugars, amino acids, and organic acids [58]. Bacteroidetes can degrade more recalcitrant carbon compounds, such as cellulose, lignin, and chitin [59].

In Nitrogen Cycle (N)

The primary element present in protein and nucleic acid structure is nitrogen; also, it is a limiting factor for improving plant biomass production. Microbes play an essential role in the nitrogen cycle through several processes nitrogen fixation, nitrification, and denitrification.

Nitrogen Fixation occurs either by the symbiotic bacteria such as *Rhizobium* and *Mesorhizobium,* which interact with the plants' roots, or by the free-living microbes such as *Azotobacter, Azospirillum, Burkholderia,* and *Clostridium.* **Nitrification** includes two steps; the first step is to oxidize ammonia into nitrite by some soil bacteria such as (*Crenarchaeum, Nitrososphaera,* or *Nitrosomonas*), and the second step is to oxidize nitrite into nitrate by (*Nitrobacter* and *Nitrospira*). **Denitrification** is the process of reducing nitrate into nitrite and nitric oxide into nitrous oxide gas or nitrogen gas by some soil microbes (*Actinobacteria, Firmicutes,* or *Proteobacteria*) [1].

Soil Quality and Agriculture Improvement

The main parameters for crop growth and production are soil quality, soil physicochemical properties, and environmental conditions. According to many studies, crop production can be increased by 40-50% with proficient crop cultivars and soil health improvement with various bio-organic inputs [60]. Microbes have a significant role in enhancing soil quality and increasing the agriculture yield [61, 62] in many ways, *i.e.*, fixing atmospheric nitrogen and soil mineral phosphates, converting insoluble elements into soluble forms so plants can obtain them easily (potassium, phosphates, nitrogen, carbon, and iron), and improving the plant's defense system against environmental stresses [13, 14], by producing enzymes, plant hormones, and organic acids.

Microbes in the soil system help to make soil firm with a good amount of organic substance [63]. The amount and value of microbial biomass are related to the available nutrient levels of soil composition. Microbes have a significant role in providing nutrients to plants by fixing nitrogen and phosphorus by symbiotic bacteria found near the plant's roots, such as (*Rhizobium* spp and *Mesorhizobium* spp, *etc.*). Free-living microbials such as *Bacillus* spp, *Pseudomonas* spp, *Thiobacillus* spp, *Achromobacter* spp, and *Azotobacter* spp [64 - 66], solubilizing and mineralizing elements, and phosphorus-solubilizing microorganisms (PSMs) such as *Pseudomonas, Bacillus, Rhizobium* and *Enterobacter*) can solubilize and mineralize phosphorus [67 - 69] by producing organic acids, then discharging the organic phosphorus by acid phosphatase. When PSM is co-inoculated with other beneficial bacteria or mycorrhizal fungi, it becomes more efficacious [70]. However, bacteria are more effective than fungi in solubilizing phosphorous [69]. Another example of solubilizing microbes is Potassium Solubilizing Bacteria (KSB) that can convert insoluble K into soluble forms so plants can obtain K easily. *Azotobacter, Acidithiobacillus, Arthrobacter, Bacillus,* and *Paenibacillus* bacteria can solubilize K minerals such as (feldspar, muscovite, biotite, mica, iolite, and orthoclase) by producing organic acids for example (oxalic, citric, tartaric and gluconic) [71 - 75]. They control the expansion of pathogenic microbes by killing harmful pathogenic microbes by some microbes such as *Penicillium* sp. and *Streptomyces* sp. that are used to produce antibiotics in soil, like penicillin and streptomycin, respectively [76].

Agricultural Improvement: Microbes perform many roles in agricultural improvement *viz*; plants nutrition, fixation of atmospheric nitrogen, enhancing plant stress tolerance, *etc*.

Plants nutrition: Some microbes' genera such as (*Azospirillum, Acinetobacter, Azotobacter, Bacillus, Bradyrhizobium, Enterobacter, Pseudomonas, Rhizobium, Streptomyces,* and *Xanthomonas*) can promote plant growth and development [77 - 80] by helping the plant in acquiring and absorbing nutrients and essential elements such as nitrogen, phosphorus, potassium, and iron [81]. Also, different algal genera from the rice fields can improve plant growth by fixing atmospheric nitrogen, such as (*Aphanocapsa, Chroococcus*, and *Phormidium*) [82].

Improves Plant Stress Tolerance

Agricultural production worldwide can be affected by biotic stresses (such as, insect, fungal diseases, bacterial diseases, and viruses diseases) and abiotic stresses (such as, drought, heat, cold, flooding, and salinity), which cause more than 40% of losses in agricultural production [83, 84]. Several microbes can provide resistance to environmental stress factors (biotic and abiotic stress) [85 - 87]. Microbes that live under extreme conditions have developed many strategies to adapt to these extreme conditions. Many biomolecules found in microbes are responsible for providing stress tolerance. For example, trehalose stabilizes cell membranes and prevents protein deformation in *E. coli*, protecting from cold stress, but in the case of yeast, it protects from osmotic [88].

Abiotic Stress

Drought Stress poses a real threat to plant growth, development, and productivity [89, 90]. Drought affects physiological and biochemical processes, mostly photosynthesis, respiration, translocation, phytohormones production, adsorption of ions, sugar, and nutrient metabolism [91, 92]. So, Inoculating *Arabidopsis* with *Paenibacillus polymyxa* improved drought tolerance [93]. Also, Timmusk *et al.* [94, 95] stated that plant tolerance to drought stress could be improved by inoculating the plant with *Paenibacillus* sp. and *Bacillus* sp. Moreover, co-inoculation *of Rhizobium tropici* and *P. polymyxa* enhanced bean plant height, shoot, and nodule number under drought stress conditions [96]. Arzanesh *et al.*

[97] and Gałązka *et al.* [98], reported that *Azospirillum brasilense, A. halopreaferanse, A. irakense,* and *A. lipoferum* have the potential to improve drought tolerance in *Triticum aestivum*. Endophytic *Bacillus subtilis* EPB5, EPB22, and EPB 31 can induce water stress-related proteins and enzymes in green gram (*Vigna radiata*) plants [99]. In addition, inoculating wheat seeds with *Azospirillum lipoferum* strains could be a practical approach to improving drought tolerance [100].

Salinity Stress is a severe problem in agriculture [101 - 103]. Presently, saline stresses affect over 20% of farmland worldwide. Thus, inoculating maize with *Azospirillum* helped in enhancing salt tolerance [104]. Also, in maize, using *Azotobacter chroococcum* inoculation improved salinity stress tolerance and increased plant biomass [105]. Sukweenadhi *et al.* [106] reported inoculating *Arabidopsis thaliana* with *Paenibacillus yonginensis* DCY84T and *Micrococcus yunnanensis* PGPB7 made it more resistant to different abiotic stresses, especially salinity stress, drought stress and heavy metal (aluminum) than control plants. Also, under salinity stress conditions inoculating *Arabidopsis* transgenic plants with *Nostoc flagelliforme* improved seed germination and shoot growth [107]. Yoolong *et al.* [108] reported that Inoculating rice with *Streptomyces venezuelae* ATCC 10712 strain promoted plant growth under salinity stress by reducing ethylene and Na^+ ion contents and accumulating proline, chlorophyll content, relative water content, malondialdehyde, and K^+ ions. Moreover, *Trichoderma longibrachiatum* enhanced salt tolerance in wheat [109].

Heat and Cold Stress (both high and low temperature) is one of the significant causes of crop losses. *Pseudomonas* AKM-P6 improved heat stress-tolerant capacity in *Sorghum* [110]. Also, Ali *et al.* [111] reported that under heat stress, inculcating wheat plants with *Pseudomonas. putida* strain AKMP7 helped improve wheat plants' survival and growth by increasing tiller and grain formation, plant biomass, and root length. Under low-temperature conditions, *Burkholderia phytofirmans* strain PsJN can promote root growth, plant biomass, and physiological activity in grapes [112].

Biotic Stress

The disease is the main threat to crop production worldwide throughout the production process, from planting to harvesting to storing crop seeds [113]. Pathogens, including viruses, mycoplasma, bacteria, fungi, and nematodes, can negatively affect crops and lead to a severe loss of crop productivity. Certain strains of microbes such as *Bacillus, Pseudomonas, Streptomyces, Azotobacter,* and *Rhizobium* play a significant role as a biocontrol agent against plant pathogens by producing antibiotics and stimulating the plant's defense system, also by promoting plant growth by increasing soil fertility which helps the plants in facing pathogens. Some pathogens include *Plasmopara, Fusarium, Pythium, Phytopthora, Alternaria, Rhizoctonia, Ampelomyces, Coniothyrium, Erwinia, Glomerella,* and *Trichoderma,* [105, 108, 114 - 116]. The *Bacillus* (RAB) sp. is considered a biological control agent against *Burkholderia glumae* and *B. gladioli* which causes bacterial panicle blight of rice [116]. Also, *Pseudomonas aeruginosa* FP6 could be used as a biological control agent against *Colletotrichum gloeosporioides* and *Rhizoctonia solani,* which is responsible for anthracnose (fruit rot) in chili [117].

Microbial Role in Causing Diseases

As microbes have beneficial effects on the environment, they also have harmful effects, which humans, animals, and plant diseases represent. A pathogen is a term for microbes that cause infectious diseases [114, 118]. Microbes cause different types of infectious diseases dependent upon the interaction between the host, pathogen, and the environment [46, 118]. Microbes such as (viruses, fungi, and bacteria) are the most common pathogens which cause plant, animal, and human diseases [118].

A virus is a parasite that replicates only inside the living cell or organisms. Virus can infect all types of life forms [118]. Human viruses such as variola virus, *Flavivirus, Coronavirus, and Paramyxovirus* cause smallpox, yellow virus, severe acute respiratory syndrome, and mumps, respectively [1, 15, 118]. Plant viruses such as *Tobacco mosaic virus, Sugarcane mosaic virus, Peanut stunt virus,* and *Citrus psorosis virus* cause Yellow streaking of leaves, Sugarcane leaf mosaic, Peanut dwarfing or stunting, and bark scaling, respectively [1].

Fungi, most plant diseases are caused by fungi (85%). Fungal pathogens are causing a severe loss in crops; for example, **Stripe (yellow) rust** is a foliar disease of wheat (*Triticum aestivum* L.) caused by *Puccinia striiformis* f. s. *tritici*, resulting in a severe loss in the highly susceptible wheat cultivars around the world [119]. Since the late 1990s and early 2000s, the yield losses reached up to 20% and 40% in Central Asian countries [119, 120]. During the 2009 and 2010 growing seasons, the losses reached up to 50% in Central and West Asia. Turkey and Syria were the most affected countries, followed by Ethiopia (45%), Uzbekistan, and Morocco (35%) [121]. The late blight of potato and tomato caused by *Phytophthora infestans* led to a vast reduction and caused the Irish potato famine during1845–1847 [122].

Wheat blast is caused by the fungal pathogen *Pyricularia graminis tritici* (*Pygt*) [123]. There has been a significant disease across the world since it was first described in the states of Paran´a, Brazil, in 1985 [121].

Additional examples of fungal pathogens including *Plasmopara viticola, Alternaria solani Fusariium oxysporum, Puccinia graminis,* and *Puccinia triticina* Eriks cause Downy mildew, Early blight, Fusarium wilt, Stem rust, and Leave rust, respectively [119, 120, 121, 124].

Bacteria can lead to some severe diseases in humans, animals, and plants [125]. In 1876, anthrax in cattle and sheep (caused by *Bacillus anthracis*) was the first bacterial disease discovered, followed by the discovery of the fire blight of pear and apple (caused by *Erwinia amylovora*) by T. J. Burrill from the University of Illinois (1877–1885) [121, 125].

CONCLUSION

Microbial species play a crucial role in the functioning of ecosystems by regulating the biogeochemical cycles of life-critical elements such as carbon (C) and nitrogen (N). They have tremendous potential for energy conversion and regeneration and are likely to be an essential component of the response to climate change in the environment.

CONSENT FOR PUBLICATION

Not applicable.

CONFLICT OF INTEREST

The author declares no conflict of interest, financial or otherwise.

ACKNOWLEDGEMENT

Declared none.

REFERENCES

[1]   Edwards SJ, Kjellerup BV. Applications of biofilms in bioremediation and biotransformation of persistent organic pollutants, pharmaceuticals/personal care products, and heavy metals. Appl Microbiol Biotechnol 2013; 97(23): 9909-21.
[http://dx.doi.org/10.1007/s00253-013-5216-z] [PMID: 24150788]

[2]   Vu B, Chen M, Crawford R, Ivanova E. Bacterial extracellular polysaccharides involved in biofilm formation. Molecules 2009; 14(7): 2535-54.
[http://dx.doi.org/10.3390/molecules14072535] [PMID: 19633622]

[3]   Banerjee A. Toxic effect and bioremediation of oil contamination in algal perspective. Bioremediation for Environmental Sustainability 2020; 283-98.

[4]   Boles BR, Thoendel M, Singh PK. Self-generated diversity produces "insurance effects" in biofilm communities. Proc Natl Acad Sci USA 2004; 101(47): 16630-5.
[http://dx.doi.org/10.1073/pnas.0407460101] [PMID: 15546998]

[5]   Ong SA, Ho LN, Wong YS, Raman K. Performance and kinetic study on bioremediation of diazo dye

(Reactive Black 5) in wastewater using spent GAC–biofilm sequencing batch reactor. Water Air Soil Pollut 2012; 223(4): 1615-23.
[http://dx.doi.org/10.1007/s11270-011-0969-4]

[6] Tribelli PM, Di Martino C, López NI, Raiger Iustman LJ. Biofilm lifestyle enhances diesel bioremediation and biosurfactant production in the Antarctic polyhydroxyalkanoate producer *Pseudomonas extremaustralis.* Biodegradation 2012; 23(5): 645-51.
[http://dx.doi.org/10.1007/s10532-012-9540-2] [PMID: 22302594]

[7] Torresi E, Gülay A, Polesel F, *et al.* Reactor staging influences microbial community composition and diversity of denitrifying MBBRs- Implications on pharmaceutical removal. Water Res 2018; 138: 333-45.
[http://dx.doi.org/10.1016/j.watres.2018.03.014] [PMID: 29635164]

[8] Kumari S, Das S. Expression of metallothionein encoding gene bmtA in biofilm-forming marine bacterium *Pseudomonas aeruginosa* N6P6 and understanding its involvement in Pb(II) resistance and bioremediation. Environ Sci Pollut Res Int 2019; 26(28): 28763-74.
[http://dx.doi.org/10.1007/s11356-019-05916-2] [PMID: 31376126]

[9] Accinelli C, Saccà ML, Mencarelli M, Vicari A. Application of bioplastic moving bed biofilm carriers for the removal of synthetic pollutants from wastewater. Bioresour Technol 2012; 120: 180-6.
[http://dx.doi.org/10.1016/j.biortech.2012.06.056] [PMID: 22797083]

[10] Stokes JD, Paton GI, Semple KT. Behaviour and assessment of bioavailability of organic contaminants in soil: relevance for risk assessment and remediation. Soil Use Manage 2005; 21(1): 475-86.
[http://dx.doi.org/10.1079/SUM2005347]

[11] Valentin L, Nousiainen A, Mikkonen A. Introduction to organic contaminants in soil: concepts and risks. Emerging Organic Contaminants in Sludges 2013; 1-29.
[http://dx.doi.org/10.1007/698_2012_208]

[12] Andreu V, Picó Y. Determination of pesticides and their degradation products in soil: critical review and comparison of methods. Trends Analyt Chem 2004; 23(10-11): 772-89.
[http://dx.doi.org/10.1016/j.trac.2004.07.008]

[13] Bermúdez-Couso A, Arias-Estévez M, Nóvoa-Muñoz JC, López-Periago E, Soto-González B, Simal-Gándara J. Seasonal distributions of fungicides in soils and sediments of a small river basin partially devoted to vineyards. Water Res 2007; 41(19): 4515-25.
[http://dx.doi.org/10.1016/j.watres.2007.06.029] [PMID: 17624393]

[14] Gennadiev AN, Tsibart AS. Pyrogenic polycyclic aromatic hydrocarbons in soils of reserved and anthropogenically modified areas: Factors and features of accumulation. Eurasian Soil Sci 2013; 46(1): 28-36.
[http://dx.doi.org/10.1134/S106422931301002X]

[15] Wang HJ, Chen HP. Understanding the recent trend of haze pollution in eastern China: roles of climate change. Atmos Chem Phys 2016; 16(6): 4205-11.
[http://dx.doi.org/10.5194/acp-16-4205-2016]

[16] Lièvremont D, Bertin PN, Lett MC. Arsenic in contaminated waters: Biogeochemical cycle, microbial metabolism and biotreatment processes. Biochimie 2009; 91(10): 1229-37.
[http://dx.doi.org/10.1016/j.biochi.2009.06.016] [PMID: 19567262]

[17] Kumar A, Bisht BS, Joshi VD, Dhewa T. Review on bioremediation of polluted environment: a management tool. Int J Environ Sci 2011; 1(6): 1079-93.

[18] Mueller JG, Cerniglia CE, Pritchard PH. Bioremediation of environments contaminated by polycyclic aromatic hydrocarbons. Biotechnol Res Series 1996; 6: 125-94.
[http://dx.doi.org/10.1017/CBO9780511608414.007]

[19] Hou D, O'Connor D, Igalavithana AD, *et al.* Metal contamination and bioremediation of agricultural soils for food safety and sustainability. Nat Rev Earth Environ 2020; 1(7): 366-81.

[http://dx.doi.org/10.1038/s43017-020-0061-y]

[20]   Hou J, Lin D, White JC, Gardea-Torresdey JL, Xing B. Joint nanotoxicology assessment provides a new strategy for developing nanoenabled bioremediation technologies. Environ Sci Technol 2019; 53(14): 7927-9.
[http://dx.doi.org/10.1021/acs.est.9b03593] [PMID: 31269395]

[21]   Kallmeyer J, Pockalny R, Adhikari RR, Smith DC, D'Hondt S. Global distribution of microbial abundance and biomass in subseafloor sediment. Proc Natl Acad Sci USA 2012; 109(40): 16213-6.
[http://dx.doi.org/10.1073/pnas.1203849109] [PMID: 22927371]

[22]   Serna-Chavez HM, Fierer N, van Bodegom PM. Global drivers and patterns of microbial abundance in soil. Glob Ecol Biogeogr 2013; 22(10): 1162-72.
[http://dx.doi.org/10.1111/geb.12070]

[23]   Anantharaman K, Brown CT, Hug LA, *et al.* Thousands of microbial genomes shed light on interconnected biogeochemical processes in an aquifer system. Nat Commun 2016; 7(1): 13219.
[http://dx.doi.org/10.1038/ncomms13219] [PMID: 27774985]

[24]   Frutos FJG, Pérez R, Escolano O, *et al.* Remediation trials for hydrocarbon-contaminated sludge from a soil washing process: Evaluation of bioremediation technologies. J Hazard Mater 2012; 199-200: 262-71.
[http://dx.doi.org/10.1016/j.jhazmat.2011.11.017] [PMID: 22118850]

[25]   Smith E, Thavamani P, Ramadass K, Naidu R, Srivastava P, Megharaj M. Remediation trials for hydrocarbon-contaminated soils in arid environments: Evaluation of bioslurry and biopiling techniques. Int Biodeterior Biodegradation 2015; 101: 56-65.
[http://dx.doi.org/10.1016/j.ibiod.2015.03.029]

[26]   Azubuike CC, Chikere CB, Okpokwasili GC. Bioremediation techniques–classification based on site of application: principles, advantages, limitations and prospects. World J Microbiol Biotechnol 2016; 32(11): 180.
[http://dx.doi.org/10.1007/s11274-016-2137-x] [PMID: 27638318]

[27]   Scow KM, Hicks KA. Natural attenuation and enhanced bioremediation of organic contaminants in groundwater. Curr Opin Biotechnol 2005; 16(3): 246-53.
[http://dx.doi.org/10.1016/j.copbio.2005.03.009] [PMID: 15961025]

[28]   García-Delgado C, Alfaro-Barta I, Eymar E. Combination of biochar amendment and mycoremediation for polycyclic aromatic hydrocarbons immobilization and biodegradation in creosote-contaminated soil. J Hazard Mater 2015; 285: 259-66.
[http://dx.doi.org/10.1016/j.jhazmat.2014.12.002] [PMID: 25506817]

[29]   Silva-Castro GA, Uad I, Gónzalez-López J, Fandiño CG, Toledo FL, Calvo C. Application of selected microbial consortia combined with inorganic and oleophilic fertilizers to recuperate oil-polluted soil using land farming technology. Clean Technol Environ Policy 2012; 14(4): 719-26.
[http://dx.doi.org/10.1007/s10098-011-0439-0]

[30]   Bhattacharya M, Guchhait S, Biswas D, Datta S. Waste lubricating oil removal in a batch reactor by mixed bacterial consortium: a kinetic study. Bioprocess Biosyst Eng 2015; 38(11): 2095-106.
[http://dx.doi.org/10.1007/s00449-015-1449-9] [PMID: 26271337]

[31]   O'Connor D, Hou D, Ok YS, Lanphear BP. The effects of iniquitous lead exposure on health. Nat Sustain 2020; 3(2): 77-9.
[http://dx.doi.org/10.1038/s41893-020-0475-z]

[32]   Kumar M, Jaiswal S, Sodhi KK, *et al.* Antibiotics bioremediation: Perspectives on its ecotoxicity and resistance. Environ Int 2019; 124: 448-61.
[http://dx.doi.org/10.1016/j.envint.2018.12.065] [PMID: 30684803]

[33]   Regenberg B, Hanghøj KE, Andersen KS, Boomsma JJ. Clonal yeast biofilms can reap competitive advantages through cell differentiation without being obligatorily multicellular. 1842.

[34]   Lohse MB, Gulati M, Johnson AD, Nobile CJ. Development and regulation of single- and multi-species Candida albicans biofilms. Nat Rev Microbiol 2018; 16(1): 19-31.
[http://dx.doi.org/10.1038/nrmicro.2017.107] [PMID: 29062072]

[35]   Banerjee D, Shivapriya PM, Gautam PK, Misra K, Sahoo AK, Samanta SK. A review on basic biology of bacterial biofilm infections and their treatments by nanotechnology-based approaches. Proc Natl Acad Sci, India, Sect B Biol Sci 2020; 90(2): 243-59.
[http://dx.doi.org/10.1007/s40011-018-01065-7]

[36]   Sharma G, Karnwal A. Biological Strategies Against Biofilms. Microbial Biotechnology: Basic Research and Applications 2020; 205-32.
[http://dx.doi.org/10.1007/978-981-15-2817-0_9]

[37]   Vasudevan R. Biofilms: microbial cities of scientific significance. J Microbiol Exp 2014; 1(3): 00014.
[http://dx.doi.org/10.15406/jmen.2014.01.00014]

[38]   Sgountzos IN, Pavlou S, Paraskeva CA, Payatakes AC. Growth kinetics of *Pseudomonas fluorescens* in sand beds during biodegradation of phenol. Biochem Eng J 2006; 30(2): 164-73.
[http://dx.doi.org/10.1016/j.bej.2006.03.005]

[39]   Jung JH, Lee SS, Shinkai S, Iwaura R, Shimizu T. Novel silica nanotubes using a library of carbohydrate gel assemblies as templates for sol-gel transcription in binary systems. Bull Korean Chem Soc 2004; 25(1): 63-8.
[http://dx.doi.org/10.5012/bkcs.2004.25.1.063]

[40]   Costerton JW, Lewandowski Z, Caldwell DE, Korber DR, Lappin-Scott HM. Microbial biofilms. Annu Rev Microbiol 1995; 49(1): 711-45.
[http://dx.doi.org/10.1146/annurev.mi.49.100195.003431] [PMID: 8561477]

[41]   Manning AJ, Kuehn MJ. Functional advantages conferred by extracellular prokaryotic membrane vesicles. J Mol Microbiol Biotechnol 2013; 23(1-2): 131-41.
[PMID: 23615201]

[42]   Beveridge TJ, Makin SA, Kadurugamuwa JL, Li Z. Interactions between biofilms and the environment. FEMS Microbiol Rev 1997; 20(3-4): 291-303.
[http://dx.doi.org/10.1111/j.1574-6976.1997.tb00315.x] [PMID: 9299708]

[43]   Mangwani N, Kumari S, Das S. Bacterial biofilms and quorum sensing: fidelity in bioremediation technology. Biotechnol Genet Eng Rev 2016; 32(1-2): 43-73.
[http://dx.doi.org/10.1080/02648725.2016.1196554] [PMID: 27320224]

[44]   Owlad M, Aroua MK, Daud WAW, Baroutian S. Removal of hexavalent chromium-contaminated water and wastewater: a review. Water Air Soil Pollut 2009; 200(1-4): 59-77.
[http://dx.doi.org/10.1007/s11270-008-9893-7]

[45]   Igiri BE, Okoduwa SIR, Idoko GO, Akabuogu EP, Adeyi AO, Ejiogu IK. Toxicity and bioremediation of heavy metals contaminated ecosystem from tannery wastewater: a review. J Toxicol 2018; 2018: 1-16.
[http://dx.doi.org/10.1155/2018/2568038] [PMID: 30363677]

[46]   Mnif I, Sahnoun R, Ellouz-Chaabouni S, Ghribi D. Application of bacterial biosurfactants for enhanced removal and biodegradation of diesel oil in soil using a newly isolated consortium. Process Saf Environ Prot 2017; 109: 72-81.
[http://dx.doi.org/10.1016/j.psep.2017.02.002]

[47]   Mallick S, Chakraborty J, Dutta TK. Role of oxygenases in guiding diverse metabolic pathways in the bacterial degradation of low-molecular-weight polycyclic aromatic hydrocarbons: A review. Crit Rev Microbiol 2011; 37(1): 64-90.
[http://dx.doi.org/10.3109/1040841X.2010.512268] [PMID: 20846026]

[48]   Cerniglia CE. Biodegradation of polycyclic aromatic hydrocarbons. Biodegradation 1992; 3(2-3): 351-68.

[http://dx.doi.org/10.1007/BF00129093]

[49]   Moody JD, Freeman JP, Fu PP, Cerniglia CE. Degradation of Benzo[*a*]pyrene by *Mycobacterium vanbaalenii* PYR-1. Appl Environ Microbiol 2004; 70(1): 340-5.
[http://dx.doi.org/10.1128/AEM.70.1.340-345.2004] [PMID: 14711661]

[50]   Foght J. Anaerobic biodegradation of aromatic hydrocarbons: pathways and prospects. Microbial Physiology 2008; 15(2-3): 93-120.
[http://dx.doi.org/10.1159/000121324] [PMID: 18685265]

[51]   Carmona M, Zamarro MT, Blázquez B, *et al.* Anaerobic catabolism of aromatic compounds: a genetic and genomic view. Microbiol Mol Biol Rev 2009; 73(1): 71-133.
[http://dx.doi.org/10.1128/MMBR.00021-08] [PMID: 19258534]

[52]   Grimm AC, Harwood CS. Chemotaxis of *Pseudomonas* spp. to the polyaromatic hydrocarbon naphthalene. Appl Environ Microbiol 1997; 63(10): 4111-5.
[http://dx.doi.org/10.1128/aem.63.10.4111-4115.1997] [PMID: 9327579]

[53]   Plósz BG, Benedetti L, Daigger GT, *et al.* Modelling micro-pollutant fate in wastewater collection and treatment systems: status and challenges. Water Sci Technol 2013; 67(1): 1-15.
[http://dx.doi.org/10.2166/wst.2012.562] [PMID: 23128615]

[54]   Nhi Cong LT, Ngoc Mai CT, Thanh VT, Nga LP, Minh NN. Application of a biofilm formed by a mixture of yeasts isolated in Vietnam to degrade aromatic hydrocarbon polluted wastewater collected from petroleum storage. Water Sci Technol 2014; 70(2): 329-36.
[http://dx.doi.org/10.2166/wst.2014.233] [PMID: 25051481]

[55]   Fagervold SK, Watts JEM, May HD, Sowers KR. Sequential reductive dechlorination of meta-chlorinated polychlorinated biphenyl congeners in sediment microcosms by two different Chloroflexi phylotypes. Appl Environ Microbiol 2005; 71(12): 8085-90.
[http://dx.doi.org/10.1128/AEM.71.12.8085-8090.2005] [PMID: 16332789]

[56]   Macedo AJ, Kuhlicke U, Neu TR, Timmis KN, Abraham WR. Three stages of a biofilm community developing at the liquid-liquid interface between polychlorinated biphenyls and water. Appl Environ Microbiol 2005; 71(11): 7301-9.
[http://dx.doi.org/10.1128/AEM.71.11.7301-7309.2005] [PMID: 16269772]

[57]   Payne RB, May HD, Sowers KR. Enhanced reductive dechlorination of polychlorinated biphenyl impacted sediment by bioaugmentation with a dehalorespiring bacterium. Environ Sci Technol 2011; 45(20): 8772-9.
[http://dx.doi.org/10.1021/es201553c] [PMID: 21902247]

[58]   Mercier A, Wille G, Michel C, *et al.* Biofilm formation *vs.* PCB adsorption on granular activated carbon in PCB-contaminated aquatic sediment. J Soils Sediments 2013; 13(4): 793-800.
[http://dx.doi.org/10.1007/s11368-012-0647-1]

[59]   Lohman K, Seigneur C. Atmospheric fate and transport of dioxins: local impacts. Chemosphere 2001; 45(2): 161-71.
[http://dx.doi.org/10.1016/S0045-6535(00)00559-2] [PMID: 11572608]

[60]   Hiraishi A, Miyakoda H, Lim BR, Hu HY, Fujie K, Suzuki J. Toward the bioremediation of dioxin-polluted soil: structural and functional analyses of *in situ* microbial populations by quinone profiling and culture-dependent methods. Appl Microbiol Biotechnol 2001; 57(1-2): 248-56.
[http://dx.doi.org/10.1007/s002530100751] [PMID: 11693929]

[61]   Yoshida S, Ogawa N, Fujii T, Tsushima S. Enhanced biofilm formation and 3-chlorobenzoate degrading activity by the bacterial consortium of *Burkholderia* sp. NK8 and *Pseudomonas aeruginosa* PAO1. J Appl Microbiol 2009; 106(3): 790-800.
[http://dx.doi.org/10.1111/j.1365-2672.2008.04027.x] [PMID: 19191976]

[62]   Löffler FE, Ritalahti KM, Zinder SH. Dehalococcoides and reductive dechlorination of chlorinated solvents. In: Stroo HF, Ed. Bioaugmentation for groundwater remediation 2013; Vol. 5: 39-88.

[http://dx.doi.org/10.1007/978-1-4614-4115-1_2]

[63]   Ziv-El MC, Rittmann BE. Systematic evaluation of nitrate and perchlorate bioreduction kinetics in groundwater using a hydrogen-based membrane biofilm reactor. Water Res 2009; 43(1): 173-81.
        [http://dx.doi.org/10.1016/j.watres.2008.09.035] [PMID: 18951606]

[64]   Conrad ME, Brodie EL, Radtke CW, *et al.* Field evidence for co-metabolism of trichloroethene stimulated by addition of electron donor to groundwater. Environ Sci Technol 2010; 44(12): 4697-704.
        [http://dx.doi.org/10.1021/es903535j] [PMID: 20476753]

[65]   Lohner ST, Becker D, Mangold KM, Tiehm A. Sequential reductive and oxidative biodegradation of chloroethenes stimulated in a coupled bioelectro-process. Environ Sci Technol 2011; 45(15): 6491-7.
        [http://dx.doi.org/10.1021/es200801r] [PMID: 21678913]

[66]   Latch DE, Packer JL, Arnold WA, McNeill K. Photochemical conversion of triclosan to 2,8-dichlorodibenzo-p-dioxin in aqueous solution. J Photochem Photobiol Chem 2003; 158(1): 63-6.
        [http://dx.doi.org/10.1016/S1010-6030(03)00103-5]

[67]   Buth JM, Grandbois M, Vikesland PJ, McNeill K, Arnold WA. Aquatic photochemistry of chlorinated triclosan derivatives: potential source of polychlorodibenzo-p-dioxins. Environ Toxicol Chem 2009; 28(12): 2555-63.
        [http://dx.doi.org/10.1897/08-490.1] [PMID: 19908930]

[68]   Heidler J, Halden RU. Mass balance assessment of triclosan removal during conventional sewage treatment. Chemosphere 2007; 66(2): 362-9.
        [http://dx.doi.org/10.1016/j.chemosphere.2006.04.066] [PMID: 16766013]

[69]   Winkler M, Lawrence JR, Neu TR. Selective degradation of ibuprofen and clofibric acid in two model river biofilm systems. Water Res 2001; 35(13): 3197-205.
        [http://dx.doi.org/10.1016/S0043-1354(01)00026-4] [PMID: 11487117]

[70]   Sui Q, Huang J, Deng S, Chen W, Yu G. Seasonal variation in the occurrence and removal of pharmaceuticals and personal care products in different biological wastewater treatment processes. Environ Sci Technol 2011; 45(8): 3341-8.
        [http://dx.doi.org/10.1021/es200248d] [PMID: 21428396]

[71]   Zwiener C, Frimmel F. Short-term tests with a pilot sewage plant and biofilm reactors for the biological degradation of the pharmaceutical compounds clofibric acid, ibuprofen, and diclofenac. Sci Total Environ 2003; 309(1-3): 201-11.
        [http://dx.doi.org/10.1016/S0048-9697(03)00002-0] [PMID: 12798104]

[72]   Onesios KM, Bouwer EJ. Biological removal of pharmaceuticals and personal care products during laboratory soil aquifer treatment simulation with different primary substrate concentrations. Water Res 2012; 46(7): 2365-75.
        [http://dx.doi.org/10.1016/j.watres.2012.02.001] [PMID: 22374299]

[73]   von Canstein H, Kelly S, Li Y, Wagner-Döbler I. Species diversity improves the efficiency of mercury-reducing biofilms under changing environmental conditions. Appl Environ Microbiol 2002; 68(6): 2829-37.
        [http://dx.doi.org/10.1128/AEM.68.6.2829-2837.2002] [PMID: 12039739]

[74]   Diels L, Spaans PH, Van Roy S, *et al.* Heavy metals removal by sand filters inoculated with metal sorbing and precipitating bacteria. Hydrometallurgy 2003; 71(1-2): 235-41.
        [http://dx.doi.org/10.1016/S0304-386X(03)00161-0]

[75]   Jong T, Parry DL. Removal of sulfate and heavy metals by sulfate reducing bacteria in short-term bench scale upflow anaerobic packed bed reactor runs. Water Res 2003; 37(14): 3379-89.
        [http://dx.doi.org/10.1016/S0043-1354(03)00165-9] [PMID: 12834731]

[76]   Itusha A, Osborne WJ, Vaithilingam M. Enhanced uptake of Cd by biofilm forming Cd resistant plant growth promoting bacteria bioaugmented to the rhizosphere of *Vetiveria zizanioides*. Int J Phytoremediation 2019; 21(5): 487-95.

[http://dx.doi.org/10.1080/15226514.2018.1537245] [PMID: 30648408]

[77] Shukla SK, Hariharan S, Rao TS. Uranium bioremediation by acid phosphatase activity of Staphylococcus aureus biofilms: Can a foe turn a friend? J Hazard Mater 2020; 384: 121316.
[http://dx.doi.org/10.1016/j.jhazmat.2019.121316] [PMID: 31607578]

[78] Yousra Turki , Mehri I, Lajnef R, *et al.* Biofilms in bioremediation and wastewater treatment: characterization of bacterial community structure and diversity during seasons in municipal wastewater treatment process. Environ Sci Pollut Res Int 2017; 24(4): 3519-30.
[http://dx.doi.org/10.1007/s11356-016-8090-2] [PMID: 27878485]

[79] Khusnuryani A, Martani E, Wibawa T, Widada J. Molecular identification of phenol-degrading and biofilm-forming bacteria from wastewater and peat soil. Indones J Biotechnol 2016; 19(2): 99-110.
[http://dx.doi.org/10.22146/ijbiotech.9299]

[80] Chakraborty J, Das S. Application of spectroscopic techniques for monitoring microbial diversity and bioremediation. Appl Spectrosc Rev 2017; 52(1): 1-38.
[http://dx.doi.org/10.1080/05704928.2016.1199028]

[81] Wolf G, Crespo JG, Reis MAM. Optical and spectroscopic methods for biofilm examination and monitoring. Rev Environ Sci Biotechnol 2002; 1(3): 227-51.
[http://dx.doi.org/10.1023/A:1021238630092]

[82] Wilson C, Lukowicz R, Merchant S, *et al.* Quantitative and qualitative assessment methods for biofilm growth: A mini-review. Res Rev J Eng Technol 2017; 6(4): 1-42.
[PMID: 30214915]

[83] May HD, Cutter LA, Miller GS, Milliken CE, Watts JEM, Sowers KR. Stimulatory and inhibitory effects of organohalides on the dehalogenating activities of PCB-dechlorinating bacterium o-17. Environ Sci Technol 2006; 40(18): 5704-9.
[http://dx.doi.org/10.1021/es052521y] [PMID: 17007129]

[84] Fagervold SK, May HD, Sowers KR. Microbial reductive dechlorination of aroclor 1260 in Baltimore harbor sediment microcosms is catalyzed by three phylotypes within the phylum Chloroflexi. Appl Environ Microbiol 2007; 73(9): 3009-18.
[http://dx.doi.org/10.1128/AEM.02958-06] [PMID: 17351091]

[85] Kjellerup BV, Sun X, Ghosh U, May HD, Sowers KR. Site-specific microbial communities in three PCB-impacted sediments are associated with different *in situ* dechlorinating activities. Environ Microbiol 2008; 10(5): 1296-309.
[http://dx.doi.org/10.1111/j.1462-2920.2007.01543.x] [PMID: 18312399]

[86] O'Neil RA, Holmes DE, Coppi MV, *et al.* Gene transcript analysis of assimilatory iron limitation in Geobacteraceae during groundwater bioremediation. Environ Microbiol 2008; 10(5): 1218-30.
[http://dx.doi.org/10.1111/j.1462-2920.2007.01537.x] [PMID: 18279349]

[87] N'Guessan AL, Elifantz H, Nevin KP, *et al.* Molecular analysis of phosphate limitation in Geobacteraceae during the bioremediation of a uranium-contaminated aquifer. ISME J 2010; 4(2): 253-66.
[http://dx.doi.org/10.1038/ismej.2009.115] [PMID: 20010635]

[88] Kye-Heon Oh , Tuovinen OH. Biodegradation of the phenoxy herbicides MCPP and 2,4-D in fixed-film column reactors. Int Biodeterior Biodegradation 1994; 33(1): 93-9.
[http://dx.doi.org/10.1016/0964-8305(94)90057-4]

[89] Zhang TC, Fu YC, Bishop PL, *et al.* Transport and biodegradation of toxic organics in biofilms. J Hazard Mater 1995; 41(2-3): 267-85.
[http://dx.doi.org/10.1016/0304-3894(94)00118-Z]

[90] Puhakka JA, Melin ES, Järvinen KT, *et al.* Fluidized-bed biofilms for chlorophenol mineralization. Water Sci Technol 1995; 31(1): 227-35.
[http://dx.doi.org/10.2166/wst.1995.0051]

[91]   Jin G, Englande AJ Jr. Carbon tetrachloride biodegradation in a fixed-biofilm reactor and its kinetic study. Water Sci Technol 1998; 38(8-9): 155-62.
[http://dx.doi.org/10.2166/wst.1998.0802]

[92]   Yamaguchi T, Ishida M, Suzuki T. Biodegradation of hydrocarbons by *Prototheca zopfii* in rotating biological contactors. Process Biochem 1999; 35(3-4): 403-9.
[http://dx.doi.org/10.1016/S0032-9592(99)00086-2]

[93]   Eriksson M, Dalhammar G, Mohn WW. Bacterial growth and biofilm production on pyrene. FEMS Microbiol Ecol 2002; 40(1): 21-7.
[http://dx.doi.org/10.1111/j.1574-6941.2002.tb00932.x] [PMID: 19709207]

[94]   Kapdan IK, Kargi F. Simultaneous biodegradation and adsorption of textile dyestuff in an activated sludge unit. Process Biochem 2002; 37(9): 973-81.
[http://dx.doi.org/10.1016/S0032-9592(01)00309-0]

[95]   Vayenas DV, Michalopoulou E, Constantinides GN, Pavlou S, Payatakes AC. Visualization experiments of biodegradation in porous media and calculation of the biodegradation rate. Adv Water Resour 2002; 25(2): 203-19.
[http://dx.doi.org/10.1016/S0309-1708(01)00023-9]

[96]   Guiot SR, Tartakovsky B, Lanthier M, *et al.* Strategies for augmenting the pentachlorophenol degradation potential of UASB anaerobic granules. Water Sci Technol 2002; 45(10): 35-41.
[http://dx.doi.org/10.2166/wst.2002.0283] [PMID: 12188570]

[97]   Chang CC, Tseng SK, Chang CC, Ho CM. Degradation of 2-chlorophenol *via* a hydrogenotrophic biofilm under different reductive conditions. Chemosphere 2004; 56(10): 989-97.
[http://dx.doi.org/10.1016/j.chemosphere.2004.04.051] [PMID: 15268966]

[98]   Kargi F, Eker S. Removal of 2,4-dichlorophenol and toxicity from synthetic wastewater in a rotating perforated tube biofilm reactor. Process Biochem 2005; 40(6): 2105-11.
[http://dx.doi.org/10.1016/j.procbio.2004.07.013]

[99]   Dasgupta D, Ghosh R, Sengupta TK. Biofilm-mediated enhanced crude oil degradation by newly isolated *pseudomonas* species. ISRN Biotechnol 2013; 2013: 1-13.
[http://dx.doi.org/10.5402/2013/250749] [PMID: 25937972]

[100]   Alessandrello MJ, Juárez Tomás MS, Raimondo EE, Vullo DL, Ferrero MA. Petroleum oil removal by immobilized bacterial cells on polyurethane foam under different temperature conditions. Mar Pollut Bull 2017; 122(1-2): 156-60.
[http://dx.doi.org/10.1016/j.marpolbul.2017.06.040] [PMID: 28641883]

[101]   Lerch TZ, Chenu C, Dignac MF, Barriuso E, Mariotti A. Biofilm *vs.* Planktonic Lifestyle: consequences for Pesticide 2, 4-D metabolism by *Cupriavidus necator* JMP134. Front Microbiol 2017; 8: 904.
[http://dx.doi.org/10.3389/fmicb.2017.00904] [PMID: 28588567]

[102]   Parellada EA, Igarza M, Isacc P, *et al.* Squamocin, an annonaceous acetogenin, enhances naphthalene degradation mediated by *Bacillus atrophaeus* CN4. Rev Argent Microbiol 2017; 49(3): 282-8.
[http://dx.doi.org/10.1016/j.ram.2017.03.004] [PMID: 28554707]

[103]   An X, Cheng Y, Huang M, *et al.* Treating organic cyanide-containing groundwater by immobilization of a nitrile-degrading bacterium with a biofilm-forming bacterium using fluidized bed reactors. Environ Pollut 2018; 237: 908-16.
[http://dx.doi.org/10.1016/j.envpol.2018.01.087] [PMID: 29551479]

[104]   Kasai Y, Kishira H, Harayama S. Bacteria belonging to the genus *cycloclasticus* play a primary role in the degradation of aromatic hydrocarbons released in a marine environment. Appl Environ Microbiol 2002; 68(11): 5625-33.
[http://dx.doi.org/10.1128/AEM.68.11.5625-5633.2002] [PMID: 12406758]

[105]   Srinath T, Verma T, Ramteke PW, Garg SK. Chromium (VI) biosorption and bioaccumulation by

chromate resistant bacteria. Chemosphere 2002; 48(4): 427-35.
[http://dx.doi.org/10.1016/S0045-6535(02)00089-9] [PMID: 12152745]

[106] Aguilar-Barajas E, Paluscio E, Cervantes C, Rensing C. Expression of chromate resistance genes from *Shewanella* sp. strain ANA-3 in *Escherichia coli*. FEMS Microbiol Lett 2008; 285(1): 97-100.
[http://dx.doi.org/10.1111/j.1574-6968.2008.01220.x] [PMID: 18537831]

[107] Davis B, Ng SP, Palombo EA, Bhave M. A Tn 5051-like mer-containing transposon identified in a heavy metal tolerant strain *Achromobacter* sp. AO22. BMC Res Notes 2009; 2(1): 1-7.

[108] Achal V, Kumari D, Pan X. Bioremediation of chromium contaminated soil by a brown-rot fungus, *Gloeophyllum sepiarium*. Research Journal of Microbiology 2011; 6(2): 166-71.
[http://dx.doi.org/10.3923/jm.2011.166.171]

[109] Kumar Ramasamy R, Congeevaram S, Thamaraiselvi K. Evaluation of isolated fungal strain from e-waste recycling facility for effective sorption of toxic heavy metal Pb (II) ions and fungal protein molecular characterization-a mycoremediation approach. Asian J Exp Biol Sci 2011; 2: 342-7.

[110] Achal V, Pan X, Fu Q, Zhang D. Biomineralization based remediation of As(III) contaminated soil by *Sporosarcina ginsengisoli*. J Hazard Mater 2012; 201-202: 178-84.
[http://dx.doi.org/10.1016/j.jhazmat.2011.11.067] [PMID: 22154871]

[111] Özdemir S, Kilinc E, Poli A, Nicolaus B, Güven K. Cd, Cu, Ni, Mn and Zn resistance and bioaccumulation by thermophilic bacteria, Geobacillus toebii subsp. decanicus and Geobacillus thermoleovorans subsp. stromboliensis. World J Microbiol Biotechnol 2012; 28(1): 155-63.
[http://dx.doi.org/10.1007/s11274-011-0804-5] [PMID: 22806791]

[112] Dash HR, Das S. Bioremediation of mercury and the importance of bacterial mer genes. Int Biodeterior Biodegradation 2012; 75: 207-13.
[http://dx.doi.org/10.1016/j.ibiod.2012.07.023]

[113] Balamurugan D, Udayasooriyan C, Kamaladevi B. Chromium (VI) reduction by *Pseudomonas putida* and *Bacillus subtilis* isolated from contaminated soils. Int J Environ Sci 2014; 5(3): 522-9.

[114] Banerjee S, Gothalwal R, Sahu PK, Sao S. Microbial observation in bioaccumulation of heavy metals from the ash dyke of thermal power plants of Chhattisgarh, India. Adv Biosci Biotechnol 2015; 6(2): 131-8.
[http://dx.doi.org/10.4236/abb.2015.62013]

[115] da Silva Folli-Pereira M, *et al.* Microorganisms as Biocontrol Agents of Pests and Diseases.Microbial BioTechnology for Sustainable Agriculture. Singapore: Springer 2022; Vol. 1: pp. 143-84.
[http://dx.doi.org/10.1007/978-981-16-4843-4_4]

[116] Goel R, *et al.* Stress-tolerant beneficial microbes for sustainable agricultural production. Microorganisms for Green Revolution. Singapore: Springer 2018; pp. 141-59.
[http://dx.doi.org/10.1007/978-981-10-7146-1_8]

[117] Sasirekha B, Srividya S. Siderophore production by Pseudomonas aeruginosa FP6, a biocontrol strain for Rhizoctonia solani and Colletotrichum gloeosporioides causing diseases in chilli. Agric Nat Resour (Bangk) 2016; 50(4): 250-6.
[http://dx.doi.org/10.1016/j.anres.2016.02.003]

[118] Clercq ED. Antivirals and antiviral strategies. Nat Rev Microbiol 2004; 2(9): 704-20.
[http://dx.doi.org/10.1038/nrmicro975] [PMID: 15372081]

[119] Shahin A, Ashmawy M, El-Orabey W, Esmail S. Yield losses in wheat caused by stripe rust (Puccinia striiformis) in Egypt. American Journal of Life Sciences 2020; 8(5): 127-34.
[http://dx.doi.org/10.11648/j.ajls.20200805.17]

[120] Afzal SN, *et al.* Assessment of yield losses caused by Puccinia striiformis triggering stripe rust in the most common wheat varieties. Pak J Bot 2007; 39(6): 2127-34.

[121] Shahin A, Ashmawy M, El-Orabey W, Esmail S. Yield losses in wheat caused by stripe rust (Puccinia

striiformis) in Egypt. American Journal of Life Sciences 2020; 8(5): 127-34.
[http://dx.doi.org/10.11648/j.ajls.20200805.17]

[122] DeJarnett Alice M. The Irish Potato Famine Fungus, Phytophthora infestans (Mont.) de Bary. Ethnobotanical Leaflets 1999; 1999(2): 5.

[123] Ceresini PC, Castroagudín VL, Rodrigues FÁ, *et al.* Wheat blast: past, present, and future. Annu Rev Phytopathol 2018; 56(1): 427-56.
[http://dx.doi.org/10.1146/annurev-phyto-080417-050036] [PMID: 29975608]

[124] Kumar A, Droby S, Eds. Food Security and Plant Disease Management. Woodhead Publishing 2020.

[125] Ellis SD, Boehm MJ, Coplin D. Bacterial diseases of plants. Agric Nat Resour (Bangk) 2008; 2008: 401-6.

CHAPTER 3

Rhizospheric Microbial Communication

Shiv Shanker Gautam[1,*], **Navneet**[2], **Neelesh Babu**[3] and **Ravindra Soni**[4]

[1] *Serve India Inter College, Roshanpur, Gadarpur, Udham Singh Nagar–262401, Uttarakhand, India*

[2] *Department of Botany and Microbiology, Gurukul Kangri University, Haridwar–249404, Uttarakhand, India*

[3] *Department of Microbiology, BFIT Group of Institutions, Chakrata Road, Suddhowala, Dehradun–248007, Uttarakhand, India*

[4] *Department of Agricultural Microbiology, College of Agriculture, Indira Gandhi Krishi Vishwavidyalaya, Raipur–492012, C.G., India*

Abstract: Rhizospheric soil is enriched with diverse microbial communities, which give rise to sophisticated plant-microbes interactions *via* chemical communication. The bacteria attain communication through quorum sensing and lead to biofilm formation, developing connections between the cell density, and altering gene expression. Such processes include diffusion and accumulation of signal molecules such as autoinducer *i.e.* acyl-homoserine lactones, Autoinducer-2 (AI-2), QS pheromone, *etc.* in the environment and trigger the expression of the gene. Due to increment in cell density, bacteria produce the substances that inhibit the growth of pathogens, fix nitrogen and optimize nodule formation. Moreover, the adaptability of microbial communities under stress conditions directly/indirectly was correlated with host plant growth. The plants and soil microorganisms equally face the abiotic stresses and may cause environmental tolerance and adaptability *via* complex physiological and cellular mechanisms. The recent knowledge of the plant-microbe relationship and their communication mechanisms can be helpful in the development and commercialization of agricultural practices to improve desired crop health and productivity under various abiotic and biotic stresses. This chapter explores such habiting microbial communications in rhizosphere attributing to soil environment in various means.

Keywords: Abiotic Stress, Acyl-Homoserine Lactones, Autoinducers, Microbial Communication, Nitrogen Fixation, Nutrient Cycle, PGPR, Pheromones, Plant - Microbes Interaction, Quorum Sensing, Rhizobia, Rhizosphere, Secondary Metabolites, Signalling Molecules, Symbiosis.

* **Corresponding author Shiv Shanker Gautam:** Serve India Inter College, Roshanpur, Gadarpur, Udham Singh Nagar – 262401, Uttarakhand, India; E mail: gautam12shiv@gmail.com

Arun Karnwal & Abdel Rahman Mohammad Said Al-Tawaha (Eds.)
All rights reserved-© 2022 Bentham Science Publishers

INTRODUCTION

The rhizosphere is a complex, nutrient-rich soil with a diverse ecosystem consisting of microbial communities, including bacteria, fungi, protists, nematodes, *etc.*, that exist around the plant roots [1]. Plant roots produce various types of primary metabolites (*i.e.*, organic acids, carbohydrates, amino acids) and secondary metabolites (*i.e.*, alkaloids, flavonoids, terpenes, glycosides, phenolics), which provide an opportunity to grow, sense, nourish, communicate, and shape the microbial life with each other [2]. Moreover, plant exudates in the rhizosphere consist of low molecular weight organic substances secreted by roots, *i.e.*, sugars, amino acids and amides, enzymes, organic acids, growth factors, *etc.*, and indigenous microbial life and their products [3, 4]. These chemical exudates chiefly participate in cellular communication *via* signaling molecules produced by plants and microbes.

In contrast, rhizospheric microbes play multiple roles, including biomass degradation, production of various types of lytic enzymes, toxins, antibiotics, siderophores, and maintenance of soil structure [5]. Doubtless, rhizospheric plant – microbes interactions are the essential factor for plant growth under various abiotic and biotic conditions [6, 7]. The communication skills mainly depend upon the various signalling molecules produced by microbial communities responsible for induced gene transcription and variation of the physiological activity of the recipient microbe [7]. There are various communication systems used by microbial species, which can be distinguished by the chemical compounds produced as signaling molecules such as N-acylhomoserine lactones (AHLs), Autoinducer-2 (AI-2), QS pheromone, *etc.* [6, 8]. These microbial products help plants combat harmful pathogenic microorganisms [9] and induce systemic resistance *via* priming [10]. The priming has been observed in some non-pathogenic rhizobacteria against fungal, bacterial, and viral infections and in some insects and nematode invasion [11, 12]. Hence, this current chapter provides plant information – microbial communications in the rhizosphere contributing to the enrichment of the soil environment in various aspects. The cellular communications and interactions are quite complex and interesting and can be understood by categorizing them into three parts including the plant root – microbes communications or *vice-versa*, and microbes – microbes communications.

Plant Root – Microbes Communication

Plant roots not only provide the soil nutrients, dissolved minerals, and water to stems and leaves but also support the development of unique soil microbiota. The existence of a microbial population in the rhizosphere tends to increase the capability of plants against the soil environment and other abiotic factors. In this scenario, the growth of a particular microbial population termed niche colonization depends upon the root exudates that mediate the interaction between the plants and microbes using several signaling molecules secreted by both microbes as well as plants [13 - 15].

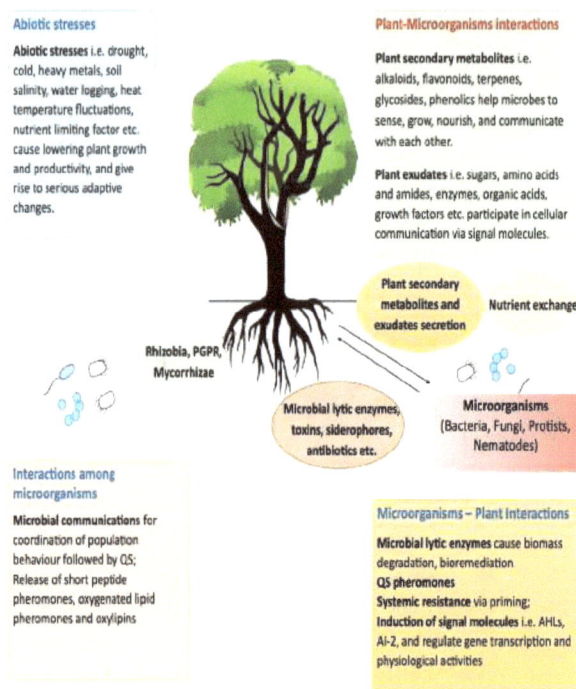

Abiotic stresses

Abiotic stresses i.e. drought, cold, heavy metals, soil salinity, water logging, heat temperature fluctuations, nutrient limiting factor etc. cause lowering plant growth and productivity, and give rise to serious adaptive changes.

Plant-Microorganisms interactions

Plant secondary metabolites i.e. alkaloids, flavonoids, terpenes, glycosides, phenolics help microbes to sense, grow, nourish, and communicate with each other.

Plant exudates i.e. sugars, amino acids and amides, enzymes, organic acids, growth factors etc. participate in cellular communication via signal molecules.

Plant secondary metabolites and exudates secretion

Nutrient exchange

Rhizobia, PGPR, Mycorrhizae

Microbial lytic enzymes, toxins, siderophores, antibiotics etc.

Microorganisms (Bacteria, Fungi, Protists, Nematodes)

Interactions among microorganisms

Microbial communications for coordination of population behaviour followed by QS; Release of short peptide pheromones, oxygenated lipid pheromones and oxylipins

Microorganisms – Plant interactions

Microbial lytic enzymes cause biomass degradation, bioremediation
QS pheromones
Systemic resistance via priming;
Induction of signal molecules i.e. AHLs, AI-2, and regulate gene transcription and physiological activities

Fig. (1). Overview of plant – microbes interactions.

Most of the studies only focused on the bacterial population, but the recent studies on fungi and protists opened future possibilities [16 - 18]. It is well known that the rhizosphere has an ample amount of nutrients that attract microbial communities. Thus, there is competition between the microbes for colonization in this highly nutritional area. The secondary metabolites produced by microbes may help in niche development and the establishment of other microbial communities either inside the roots or in the rhizosphere [19 - 22]. There are several compounds released by the microbes that include several lytic enzymes, toxins, antibiotics, siderophores, *etc*. (Fig. **1**). Various genes and gene clusters are involved in the production of these compounds such as traR/traI (pTi), braR/braI, RpfB, RpfF, pqsABCDE, pqsH, cinR/cinI, PhcB, SpeA, SpeB, *etc.* [23 - 32].

Plant Root – Bacteria Communication

Over the past few years, the importance of competitiveness, the bacteria's survival, and their communication extent have been noticed so far [7]. There is no doubt that bacterial communication is one of the essential factors for plant growth in the rhizosphere [6, 7]. The bacterial associations with plant roots may act as a plant growth-promoting agent, induce resistance, and suppress disease. Such bacterial communities are found in soil as free-living, symbiotic associations with various non-woody and woody plants, cyanobacteria and legumes [33]. They belong to many genera such as *Actinomycetes, Agrobacterium, Alcaligenes, Arthrobacter, Bacillus, Flavobacterium, Micromonospora, Nocardia, Pseudomonas, Streptomyces, etc.* [34].

Bacterial communication mainly depends upon the microbial communities' diffusion and synthesis of signal molecules. Because these molecules can induce specific responses in the gene transcription that ultimately change recipient microbes' physiological activity [7], the signal molecules can also change metabolic control and propagate certain virulence factors [35 - 38]. Regulation of such extracellular signaling molecules with respect to population density depends upon the stimulus and response termed quorum sensing (QS) [7, 35, 39, 40]. Gram-positive and Gram-negative bacteria both use the QS. Gram-positive bacteria use secreted oligopeptides and two-component systems consisting of membrane-bound sensor kinase receptors and cytoplasmic transcription factors, which alter the gene expression [41]. In comparison, Gram-negative bacteria synthesize AHLs (Fig. **2**), which act as diffusible communication signal molecules and cause discrete physiological processes in a cell-density-dependent manner [42]. Eventually, myxobacteria, a large taxon of gram-negative bacteria that live predominantly in the soil, represent a unique cellular model system to understand bacterial cell-to-cell recognition. Myxobacteria participate in the mutual exchange of outer membrane (OM) lipids and proteins *via* mediating TraA and TraB proteins without the exchange of cytoplasmic or DNA material [43, 44]. The OM exchange causes multiple roles under phenotypic and behavioral changes of bacteria, including the role of cell surface receptor, regulation of antagonistic interactions with other bacteria and fungi, specific population behavior by allele-specific mechanism, *etc.* [43, 45].

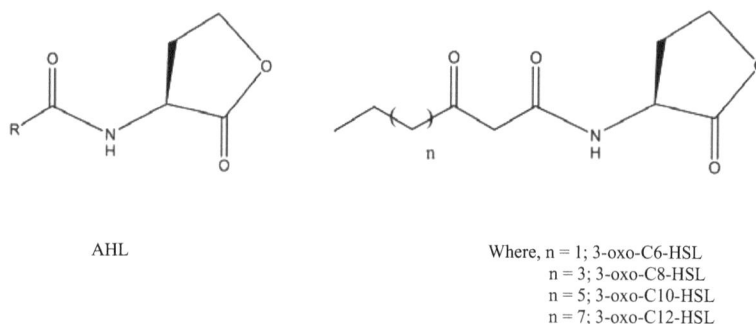

AHL

Where, n = 1; 3-oxo-C6-HSL
n = 3; 3-oxo-C8-HSL
n = 5; 3-oxo-C10-HSL
n = 7; 3-oxo-C12-HSL

Fig. (2). Structure of *N*-acylated homoserine lactone (AHL) and native AHLs molecules.

The signal-mediated communications are not restricted to similar prokaryotic organisms. These signals can be interrupted by non-related prokaryotes. This can also be used as a competitive advantage for the signal producer, who can alter the behavior of non-related species [7]. Various communication systems used by microbial species can be differentiated by the chemical compounds produced as signaling molecules (Table **1**). These signaling molecules include AHL, Autoinducer-2 (AI-2), QS pheromone, *etc.* [6, 8].

Table 1. List of signal molecules of rhizospheric bacteria.

S. No.	Signaling Molecule	Bacterial Organisms	Gene/ Protein	Role(s)	Refs.
1.	3-oxo-C8-HSL	*Agrobacterium tumefaciens, Rhizobium* sp.,	traR/traI	Autoinducer	[24, 25, 46]
2.	*N*-3-oxo-octanoyl-homoserine lactone (3OC8-HSL)	*Pseudomonas syringae*	-	Mediates priming of resistance	[47]
3.	Acyl homoserine lactones (AHLs)	*Proteobacteria, Bradyrhizobium japonicum, B. elkanii*	LuxI	Quorum sensing gene-regulatory processes	[48 - 51]
4.	Uncharacterized, non-AHL molecule	*Mesorhizobium huakuii, P. aeruginosa*	-	Autoinducer for quorum sensing	[52, 53]
5.	Cinnamoyl-HSL	*Bradyrhizobium* sp.	braR/braI	Quorum sensing exhibits plant-growth promoting activity	[26, 54]

(Table 1) cont.....

S. No.	Signaling Molecule	Bacterial Organisms	Gene/ Protein	Role(s)	Refs.
6.	3-oxo-C6-HSL	*Mesorhizobium loti*	traR/traI1/traI2	Autoinducer, significant role in quorum sensing system	[28, 55, 56]
7.	3-OH-(slc)-HSL	*Rhizobium etli*	cinR/cinI	Stimulates surface movements of cinI and cinR mutants	[30, 57]
8.	3-oxo-C12-HSL	*Burkholderia kururiensis*	braR/braI	Enhances the activity of superoxide dismutase, peroxidase and catalase under salt stress condition	[58]
9.	3-oxo-C6-HSL, 3-oxo--8-HSL, 3-oxo-C6-HSL, 3-ox--C8-HSL	*Burkholderia unamae, B. xenovorans*	braR/braI	-	[59]
10.	Bradyoxetin	*Bradyrhizobium japonicum*	-	Nodulating gene regulation in a population density dependent manner	[49, 60, 61]
11.	Linear oligopeptides	*Bacillus subtilis*	-	Extracellular signaling peptides regulate sporulation, genetic competence, and quorum responses	[62]
12.	Cyclic peptides	Gram-positive Bacteria	-	Antimicrobial properties	[63]
13.	Furanosyl diester (boron); AI-2	Diverse taxa (except α-*Proteobacter*)	LuxS AI-2 synthase	Universal signal for interspecies communication	[64, 65]
14.	cis-11-Methyl-2-dodecenoic acid	*Xanthomonas* spp.	RpfB, RpfF	Diffusible signal factor involved in synchronization of virulence gene expression and biofilm dispersal	[27, 66]
15.	4-Hydroxy-2-alkyl quinolines (HAQs)	*Pseudomonas* spp.	pqsABCDE, pqsH	Autoinducer, antimicrobial activities	[29, 67]
16.	Palmitic acid methyl esters	*Ralstonia solanacearum*	PhcB	Inhibit phagocytosis and decrease cell viability	[68]

(Table 1) cont.....

S. No.	Signaling Molecule	Bacterial Organisms	Gene/ Protein	Role(s)	Refs.
17.	3-OH-C14:1-HSL	*Rhizobium leguminosarum*	cinR/cinI	Inhibiting the growth of some strains of *Rhizobium leguminosarum* bv. *viciae*	[69, 70]
18.	A-signal	*Myxococcus xanthus*	AsgAB-dependent protease	Helps to assess starvation and induce first stage of aggregation	[71]
19.	C-signal	*Myxococcus xanthus*	-	Helps to pattern cell movement and shape the fruiting body of *M. xanthus*	[71]

Miller and Bassler [40] have suggested three basic models of quorum sensing utilized by bacterial species. First, LuxI synthesizes autoinducer AHL in the cytoplasm, which is passively diffused through the cell membrane of bacteria and gathers in intracellular as well as extracellular cell densities. The luxR-type protein is bound to AHL when the ample amount of AHL concentration is achieved to form a LuxR-AHL complex that stimulates quorum-sensing regulated targeted genes, which promotes transcription. Massively, gram-negative bacterial species secrete AHL with only a difference in their acyl side chain (Fig. **2**). Moreover, every LuxR protein binds only to their respective AHL autoinducer molecules as they are selective. In the second model, the signal molecules actively interact with external membrane-bound protein sensors as they are exported outside the cell. The phosphorylation cascades generate signal transduction; each sensor protein is highly selective for the particular peptide signals. Like gram-negative bacteria, gram-positive bacteria can also use more than one or two autoinducers. Among the peptide inducers, few act from the outside while others evoke the particular set of gene expression changes from the outside and transport them back inside the cell. In the third model, bioluminescence has been explained, which is controlled *via* quorum sensing. *Vibrio harveyi* produces autoinducer HAI-1 and AI-2, where HAI-1 is similar to AHL, the autoinducer of gram-negative bacteria, perhaps independent of an enzyme similar to LuxI. On the other hand, AI-2 is a furanosyl borate diester. Both signal molecules generate signal transduction through phosphorylation cascade similar to a Gram-positive cascade [72].

In *Agrobacterium*, the conjugal transfer of nopaline-type *Agrobacterium* Ti plasmid pTiC58 is regulated by a transcriptional activator, TraR, and a diffusible signal molecule, the conjugation factor (CF). CF is a member of a family of substituted homoserine lactones (HSLs) that act as co-inducers for regulating gene expression in Gram-negative bacteria *via* autoinduction [25]. *N*-(3-oxo-hexano-l)-L-homoserine lactone (3-oxo-C6-HSL) belongs to AHLs and has been determined as QS signaling molecule in *Serratia proteamaculans* B5a [73]. While, *Bradyrhizobium japonicum*, a symbiotic nodulating bacterium of soybean, consists of nodulation genes (nod, nol, noe) induced in the presence of plant-produced isoflavones. The expression of a nod gene is population density-dependent and mediated by an extracellular secreted factor (CDF), the novel signaling molecule designated bradyoxetin. The expression of bradyoxetin is iron-regulated and mainly produced under low iron conditions [60, 61]. Many gram-negative bacteria synthesize AHLs which help them in quorum sensing *via* the intercellular signaling mechanism responsible for activating virulent and biofilm lifestyles [48, 74].

Plant Root – Fungi Communication

The plant-fungi interactions range from pathogenicity to mutual symbiosis. Plants secrete different molecules such as phytohormones, polymers, phytochemicals, and primary and secondary metabolites. All of these play critical roles in plant-fungi interactions, and may be beneficial or deleterious [75]. Fungi use chemical signals to regulate sexual reproduction, followed by development. Usually, cyclic AMP molecules play a significant role in cell-to-cell signaling, whereas *Ustilago maydis*, the prevalent maize pathogen employs short peptide pheromones [76]. The establishment of communication between the mycorrhizae and their host usually occurs through the Myc factors or chitooligosaccharides, which are well-known plant receptors [77]. Similar to the bacterial species, most fungi effuse volatile organic compounds (VOCs) specific to each species. Oxygenated lipid pheromones and oxylipins have properties to regulate the development and pathogenesis of fungal species [78]. Unlike these, other fungal species' direct antimicrobial active VOCs might affect the phytobiome [79]. The signal molecules produced by the fungal species are prone to obstruction by other members of the phytobiome. Oxylipins, as produced by various plants, can mimic those fungal species and hence regulate the development of fungi and toxin production during pathogenesis [80]. Likewise, mimicry of fungal VOCs by plant species may insist on pollination by insects [81]. Most of the fungal VOCs can interfere with the production of exopolysaccharides of rhizospheric bacteria and may also affect their motility [82]. These fungal VOC can provide resistance to the plant species and suppress their growth, *i.e.*, *Laccaria bicolor* [83, 84].

DNA-based phylogeny suggested that arbuscular mycorrhizal fungi (AMF) have had a close interaction with plants for over 400 million years [85]. AMF has been considered to induce a signal transduction process in plant root cells that overlap with the root-nodule symbiosis and involves the induction and decoding of calcium signature [85, 86]. AMF and PGPR both work mutually in soil and plant tissues *via* increased nutrition, hyphal permeability in plant roots, rhizospheric microbial communities survivability, and protection by biotic and abiotic stresses [87]. Moreover, the rhizosphere signaling between plant roots and fungi is essential to trigger the immune response. In this context, *Trichoderma* spp. secrete hydrophobin, swollenins, TasHyd1, Qid74, and some cysteine-rich proteins, which play an essential role in plant root-fungi interaction [88]. *Trichoderma* spp. also produces expansin-like proteins with cellulose-binding modules and endopolygalacturonase to facilitate root penetration [89, 90]. Ca^{2+} molecules produced by AMF participate in regulating nodulation gene expression [86]. γ-Butyrolactones are produced by *Streptomyces* spp., which have regulatory mechanisms and act as hormone-like signal molecules [91, 92].

In reverse, the host plants also produce some chemical compounds and metabolites, which play a vital role in the establishment of the mutual symbiotic association. Strigolactones (SLs) are a group of chemical compounds that are produced by plant roots. SLs induces several responses in AMF, such as spore germination, hyphal branching, mitochondrial metabolism, transcriptional reprogramming, and production of chitin oligosaccharides. While in reverse AMF stimulates early symbiotic responses in the host plant [93]. The role of plant secondary metabolites, *i.e.*, alkaloids, flavonoids, terpenes, and glucosinolates, have been well documented in mediating below-ground interactions with plants and microbial communities [94]. Alkaloids are known as the most defensive compounds, phytotoxic, and consist of narcotic effects [95]. Flavonoids are defensive and signaling molecules that promote symbiosis, especially between legumes and rhizobia, antimicrobial, and interact with Ca^{2+} channels in the cell membrane, causing cytotoxic influx [96 - 98]. Terpenes are defensive and signaling compounds, antimicrobial, antioxidant, and phytotoxic [99, 100]. While, glucosinolates are highly reactive compounds, inhibit cytochrome P_{450}, and induce cellular apoptosis [101]. So, the plant-derived secondary metabolites not only promote the growth of neighboring plants, provide resistance against biotic and abiotic stresses, and are even helpful in balancing soil microflora.

Plant – Microbes' Nutrient Exchange

As we know, the plant provides nourishment to the rhizospheric microbial species by the supply of vital carbon sources, amino acids, sugars, *etc.*, as crucial molecules. New methodologies revealed molecular aspects and the complex mixture of metabolites in root exudates. Moreover, the particular type of metabolites present in the root exudates can inhibit or grow certain species of bacteria through biochemical mechanisms [102]. Three primary mechanisms explain the enhancement of plant growth through microbial activity, including manipulation in the signal molecules of the plants, inhibition of the pathogens, and increment of nutrient availability in the rhizosphere [103 - 105].

Several nutrients such as nitrogen, phosphorus, and sulfur are bound in different organic substances. Thus, unavailable or minimal such nutrients can happen to the plants. To access such vital nutrients, plants primarily depend on the soil microflora. These microbial species tend to depolymerize and mineralize organic forms of nitrogen, phosphorus, and sulfur [106, 107], and these nutrients are liberated in the soil in the inorganic form with some ionic species such as sulphate, phosphate, ammonium nitrate, *etc.* [103]. In general, these nutrients work as crucial ingredients for the development of plant growth and development [108]. Following the mutualistic exchange of nutrients between the plants and microbial species, carbon priming nutrients are considered. The plants with *N*-limit increase the release of metabolites from the roots to the rhizosphere, which arbitrates mobilization of microbial species that brings soil nitrogen into the bioavailable form [102]. The beneficial interactions between plants and microbes are briefly described here.

Plant Growth-Promoting Rhizobacteria (PGPR)

The symbiotic relationship between *Rhizobium* spp. and leguminous plants is one of the most important mutualistic relationships. The bacteria belong to the class Alphaproteobacteria, which are closely related to the genus *Rhizobium*, *Azorhizobium*, *Bradyrhizobium*, *Mesorhizobium*, and *Sinorhizobium*. These nitrogen-fixing bacteria are collectively called rhizobia [109]. These are responsible for the formation of nodules on roots. The nod genes are responsible for the nodule formation, and nif genes for the fixation of atmospheric nitrogen are often present in Sym plasmid. The nod gene is often transferred among several nodule formation bacteria, including α, β, and γ-Proteobacteria [109, 110].

The roots of the plants produce Phyto-phenolic compounds, *i.e.*, flavonoids, which tend to attract rhizobia. Remarkably, these bacteria bind to carbohydrate-binding proteins (lectins) on the root hairs through sugar residue. Eventually, these bacteria enter the root hairs and start producing enzymes such as cellulose and polygalacturonase that soften the cell wall of the root hairs, followed by the production of nod factors, most of the time lipo-chitooligosaccharides that cause curling of the root hairs. With the curling of root hairs, an infection thread is formed because of the invagination of the cell wall and the plasma membrane of the plant cell with the production of polysaccharides by bacteria. Further, the multiplication of the bacteria brings infection thread down the root hairs *via* the epidermis, cortex, and cortical cells to the cytoplasmic bridges until the bacteria are transported to the cortical cells that cause the nodulation [111].

Similarly, bacteria also required plant homocitrate to form the FeMo (iron-molybdenum) complex, which works as the cofactor for the nitrogenase enzyme. The regulation of O_2 causes inhibition of nitrogenase enzyme that the plant can regulate. The host plant provides energy to the bacteroids through organic acids [109].

Several types of bacterial species surround the rhizospheric region and contribute to the growth and development of the plant. These bacterial species are termed Plant Growth Promoting Rhizobia (PGPR). PGPR not only promotes plant growth but even suppresses the root pathogens. PGPR has benefited the plants as they improve the availability of nutrients, reduce stress, and protects against pathogens. PGPR includes the association of the nitrogen-fixing bacteria that uptake energy forms the root exudates and eliminates free nitrogen, which the plants utilize. These nitrogen fixers include *Acetobacter diazotrophicus*, *Herbaspirillum seropedicae*, *Azoarcus* spp., *Azotobacter* spp., *Azospirillum* spp. *etc.* These bacterial species, somewhat similar to the mutualistic Nitrogen fixers, colonize the root hairs and actively participate in the synthesis of phytohormones, including IAA, auxin, gibberellin, and cytokinin [112].

Mycorrhizae

The association between fungi and roots of higher plants is termed mycorrhizae. It is well - a studied root - fungal mutual symbiotic association. Based on fungal hyphae arrangements, these associations can be classified into seven broad types: arbuscular, ectomycorrhiza, ecto-endomycorrhiza, arbutoid, monotropoid, ericoid, and orchidaceous [113 - 115]. Brundrett [113] has described the different phases of mycorrhizae associations such as free-living, endophytes, exploitative, balanced, antagonist, and Necrotrophic. The development of mycorrhizal associations depends on the level of auxin secretion produced by associated fungi and the sensitivity of plant roots to auxin. While endogenous or induced associations depend on phosphate level in the substrate [114].

The mycorrhizae provide bi-directional movement of nutrients, carbon (C) from plants to fungi, and soil nutrients from fungi to host plants. The AM fungi-plant association is one typical example of the mutual benefit. In this, the host plant provides carbon from fixed photosynthates, and in contrast, fungi uptake the phosphate and other mineral nutrients from the soil, which are available for the plant [116]. Such a mutual relationship is benefitted by the extension of the surface of roots for the uptake of nutrition and protects the plant from pathogens and harsh conditions, *i.e.*, stress, heavy metal toxicity, *etc.* [117].

Additionally, root - fungi interactions could be parasitic or mutualistic, generally termed endophytes. Mainly endophytes belong to the phyla Basidiomycota and Ascomycota. The most frequently studied endophyte is dark septate fungi which are named because of the presence of dark pigmentation and cross walls between the hyphal cells, *e.g.*, *Phialocephala fortinii.* They vary from plant to plant and may cause both positive and adverse effects on the growth of the plants [118]. The endophytes can transfer essential soil nutrients such as carbon, nitrogen, phosphorus, iron, copper, zinc, *etc.*, to their host plants [87, 119]. Eventually, the endophytic fungi can thrive under harsh conditions *via* decomposing plant organic material, which provides nutrition to the host plant [119].

Abiotic Stresses and Soil Microbiota

The survivability of microbial communities under stress conditions was directly/indirectly correlated with plant growth. Plants and soil microorganisms both face diverse abiotic stresses like drought, variation in precipitation rate, humidity, temperature, soil pH, soil types, soil acidity and alkalinity, nutrient availability and starvatio [120 - 122], which give rise to environmental tolerance, involvement of multiple signaling pathways and adaptability *via* complex cellular mechanisms [123]. The microbial adaptive skills may consist of maintaining membrane fluidity, change in the metabolic rate, cellular processes and protein synthesis, expression of housekeeping genes, cold-shock or heat-shock proteins, *etc.* Induced systemic tolerance (IST) is another microbial-mediated induction of abiotic responses. Various microbial genera, *i.e.*, *Azotobacter, Pseudomonas, Azospirillum, Bacillus, Bradyrhizobium, cyanobacteria, Trichoderma,* and *Methylobacterium,* have been reported in plant growth promotion and mitigation of multiple kinds of abiotic stresses [123]. Thus, stress signaling helps in regulating protein and gene expression to maintain cellular stability and homeostasis under stress conditions [124].

The plant growth-promoting bacteria (PGPB) play an essential role in plant growth under various environmental stress conditions. The PGPB induces systemic resistance (ISR) as a plant defense mechanism under biotic and abiotic stresses [125]. Furthermore, nonsporulating bacteria can activate the viable but nonculturable state under abiotic stresses. Thus, environmental stress and low N_2 in the soil are also responsible for stimulating N_2 fixation [126]. Earlier studies for metabolic responses of Himalayan cold-adapted diazotrophs *Pseudomonas palleroniana* N26 have investigated nitrogen deficiency in the cold niche. Under the cold condition, expression of nifA, nifL, nifH, nifB, nifD, nifK, and cowN of the nitrogenase system has been observed [127]. Under cold stress, cellular standard protein products are decreased. In contrast, some unique protein products are increased until adaptation in certain organisms. Such a cold-induced homologous class of proteins is known as cold shock proteins (Csps) [128]. Csps are multifunctional RNA/DNA binding proteins with the presence of one or more cold shock domains [129]. Csps has been reported in various gram-positive and gram-negative bacteria, including *E. coli* [130], *Bacillus* spp [131], *Thermotoga* spp [132], *Arthrobacter* spp., *Streptococcus* spp., *Listeria* sp., and *Pseudomonas* spp [128], respectively.

Impact of Microbial Communications on Plant Health

The rhizospheric microbial–plant interactions have long been of interest that allows for steady development of the agricultural sector. It is well known how rhizosphere microbial communities play an essential role in plant health and productivity [133]. The microorganisms stimulate plant growth in various means and cause prevention of diseases, development of adaptive responses towards abiotic stresses, improvement of soil nutrient exchange, production of antibiotics, toxins, siderophores, lytic enzymes, solubilized phosphate, *etc.* [134], which make them a potential candidate for the development of sustainable agriculture. Microorganisms produce various phytohormones, small-signal molecules, and volatile compounds that directly/indirectly stimulate plant growth or immunity [135]. Further, such responses are helpful in the development of adaptive skills in a harsh environment.

Bacteria secrete various secondary metabolites which can act as bioregulators, enhance plant growth and control phytopathogenic attack. These microbial metabolites such as β-1,3-glucanase [136, 137], ACC-deaminase [138], chitinase [137], HCN [139, 140], phenazines [141, 142], pyrrolnitrin [143, 144], 2,4-diacetylphloroglucinol [145], pyoluteorin [146], viscosinamide, tensin, Amphisin [147], pyochelin [148], tetracenomycin [149], dialkylresorcinols [150], rhizoxins [151, 152], mupirocin [153, 154], oxyvinylglycines [155], orfamide A and H [156 - 159], phenazine-1-carboxylic acid [160, 161], furanomycin [162, 163], brabantamide A [164], obafluorin [165], eruginaldehyde [166], safracins [167], syringomycins SP22 or SP25 [168, 169], tabtoxin [168, 169], syringopeptins [170, 171], andrimid [172, 173], kalimantacin [174], *etc.* exhibit various properties like antimicrobial activity, insecticidal properties, mobilization of elements, elicit plant defense system, and act as biosurfactants.

Moreover, multi-omics technologies and microbial products' application as bio inoculants have revolutionized the agriculture sector in the last few decades. The preparation of bio inoculants depends upon selecting beneficial microbial strains for consortia development, suitable carriers and adjuvants, and compatibility with laboratory trials and field conditions [175]. Such bio formulations provide a protective and nutritive platform for selected microbes under various conditions.

Rhizospheric improvement is another way to maintain the soil nutrients by regulating root morphological and physiological traits, optimizing nutrient supply in the root zone, modulation of phytohormone and biocontrol activity, and deploying rhizosphere processes and interactions [176]. Similarly, plant engineering has provided the scenario for developing genetically modified crops to cope with abiotic stresses. Such crops have noticed higher productivity, quality, and yield [177, 178].

CONCLUSION

The plant-microbe relationship involves multiple events and coordination of signaling molecules. Plant and microorganisms both work simultaneously for their survival, growth, and development. The rhizospheric microorganisms are responsible for stimulating plant growth in various means and helping them in the prevention of diseases, development of adaptive responses towards abiotic stresses, and improvement of soil nutrient exchange. While in contrast, plants provide a wide range of organic compounds and metabolites which act as signaling molecules for the microbial population. These signaling molecules elicit the gene expression and metabolic pathways of rhizospheric microorganisms, which may lead to their phenotypic adaptation to their abiotic and biotic environments. Moreover, rhizospheric bacteria and fungi correspondingly enhance soil nutrients and plant growth by coordinating each other. Thus, understanding plant-microbe interactions is reasonably necessary to significantly provide a rational scenario for improving rhizosphere soil, crop health, and productivity for the fulfillment of future demands.

CONSENT FOR PUBLICATION

Not applicable.

CONFLICT OF INTEREST

The author declares no conflict of interest, financial or otherwise.

ACKNOWLEDGEMENT

Declared none.

REFERENCES

[1]   Olanrewaju OS, Ayangbenro AS, Glick BR, Babalola OO. Plant health: feedback effect of root exudates-rhizobiome interactions. Appl Microbiol Biotechnol 2019; 103(3): 1155-66.
[http://dx.doi.org/10.1007/s00253-018-9556-6] [PMID: 30570692]

[2]   Braga RM, Dourado MN, Araújo WL. Microbial interactions: ecology in a molecular perspective. Braz J Microbiol 2016; 47 (Suppl. 1): 86-98.

[http://dx.doi.org/10.1016/j.bjm.2016.10.005] [PMID: 27825606]

[3]   Canarini A, Kaiser C, Merchant A, Richter A, Wanek W. Root Exudation of Primary Metabolites: Mechanisms and their roles in plant responses to environmental stimuli. Front Plant Sci 2019; 10: 157.
[http://dx.doi.org/10.3389/fpls.2019.00157] [PMID: 30881364]

[4]   Hassan M, McInroy J, Kloepper J. The interactions of rhizodeposits with plant growth-promoting rhizobacteria in the rhizosphere: a review. Agriculture 2019; 9(7): 142.
[http://dx.doi.org/10.3390/agriculture9070142]

[5]   Furtak K, Gajda AM. Activity and variety of soil microorganisms depending on the diversity of the soil tillage system. In: de Oliveira AB, Ed. Sustainability of Agroecosystems. 2018.
[http://dx.doi.org/10.5772/intechopen.72966]

[6]   Yajima A. Recent progress in the chemistry and chemical biology of microbial signaling molecules: quorum-sensing pheromones and microbial hormones. Tetrahedron Lett 2014; 55(17): 2773-80.
[http://dx.doi.org/10.1016/j.tetlet.2014.03.051]

[7]   Atkinson S, Williams P. Quorum sensing and social networking in the microbial world. J R Soc Interface 2009; 6(40): 959-78.
[http://dx.doi.org/10.1098/rsif.2009.0203] [PMID: 19674996]

[8]   Lareen A, Burton F, Schäfer P. Plant root-microbe communication in shaping root microbiomes. Plant Mol Biol 2016; 90(6): 575-87.
[http://dx.doi.org/10.1007/s11103-015-0417-8] [PMID: 26729479]

[9]   Venturi V, Keel C. Signaling in the Rhizosphere. Trends Plant Sci 2016; 21(3): 187-98.
[http://dx.doi.org/10.1016/j.tplants.2016.01.005] [PMID: 26832945]

[10]   Serteyn L, Quaghebeur C, Ongena M, *et al.* Induced systemic resistance by a plant growth-promoting Rhizobacterium impacts development and feeding behavior of aphids. Insects 2020; 11(4): 234.
[http://dx.doi.org/10.3390/insects11040234] [PMID: 32276327]

[11]   Ramamoorthy V, Viswanathan R, Raguchander T, Prakasam V, Samiyappan R. Induction of systemic resistance by plant growth promoting rhizobacteria in crop plants against pests and diseases. Crop Prot 2001; 20(1): 1-11.
[http://dx.doi.org/10.1016/S0261-2194(00)00056-9]

[12]   Walters D, Heil M. Costs and trade-offs associated with induced resistance. Physiol Mol Plant Pathol 2007; 71(1-3): 3-17.
[http://dx.doi.org/10.1016/j.pmpp.2007.09.008]

[13]   Bulgarelli D, Rott M, Schlaeppi K, *et al.* Revealing structure and assembly cues for Arabidopsis root-inhabiting bacterial microbiota. Nature 2012; 488(7409): 91-5.
[http://dx.doi.org/10.1038/nature11336] [PMID: 22859207]

[14]   Lundberg DS, Lebeis SL, Paredes SH, *et al.* Defining the core Arabidopsis thaliana root microbiome. Nature 2012; 488(7409): 86-90.
[http://dx.doi.org/10.1038/nature11237] [PMID: 22859206]

[15]   Bai Y, Müller DB, Srinivas G, *et al.* Functional overlap of the Arabidopsis leaf and root microbiota. Nature 2015; 528(7582): 364-9.
[http://dx.doi.org/10.1038/nature16192] [PMID: 26633631]

[16]   Tedersoo L, Bahram M, Põlme S, *et al.* Global diversity and geography of soil fungi. Science 2014; 346(6213): 1256688.
[http://dx.doi.org/10.1126/science.1256688] [PMID: 25430773]

[17]   Geisen S, Tveit AT, Clark IM, *et al.* Metatranscriptomic census of active protists in soils. ISME J 2015; 9(10): 2178-90.
[http://dx.doi.org/10.1038/ismej.2015.30] [PMID: 25822483]

[18]   Heijden MGA, Martin FM, Selosse MA, Sanders IR. Mycorrhizal ecology and evolution: the past, the

present, and the future. New Phytol 2015; 205(4): 1406-23.
[http://dx.doi.org/10.1111/nph.13288] [PMID: 25639293]

[19]   Thomashow LS, Weller DM. Role of a phenazine antibiotic from *Pseudomonas fluorescens* in biological control of *Gaeumannomyces graminis* var. *tritici*. J Bacteriol 1988; 170(8): 3499-508.
[http://dx.doi.org/10.1128/jb.170.8.3499-3508.1988] [PMID: 2841289]

[20]   van Loon LC, Bakker PAHM. Induced systemic resistance as a mechanism of disease suppression by rhizobacteria. PGPR: Biocontrol and biofertilization. Dordrecht: Springer 2006; pp. 39-66.

[21]   Pierson LS III, Pierson EA. Metabolism and function of phenazines in bacteria: impacts on the behavior of bacteria in the environment and biotechnological processes. Appl Microbiol Biotechnol 2010; 86(6): 1659-70.
[http://dx.doi.org/10.1007/s00253-010-2509-3] [PMID: 20352425]

[22]   Kim YC, Leveau J, McSpadden Gardener BB, Pierson EA, Pierson LS III, Ryu CM. The multifactorial basis for plant health promotion by plant-associated bacteria. Appl Environ Microbiol 2011; 77(5): 1548-55.
[http://dx.doi.org/10.1128/AEM.01867-10] [PMID: 21216911]

[23]   Chen XH, Koumoutsi A, Scholz R, *et al.* Comparative analysis of the complete genome sequence of the plant growth–promoting bacterium *Bacillus amyloliquefaciens* FZB42. Nat Biotechnol 2007; 25(9): 1007-14.
[http://dx.doi.org/10.1038/nbt1325] [PMID: 17704766]

[24]   Piper KR, von Bodman SB, Farrand SK. Conjugation factor of *Agrobacterium tumefaciens* regulates Ti plasmid transfer by autoinduction. Nature 1993; 362(6419): 448-50.
[http://dx.doi.org/10.1038/362448a0] [PMID: 8464476]

[25]   Hwang I, Li PL, Zhang L, *et al.* TraI, a LuxI homologue, is responsible for production of conjugation factor, the Ti plasmid *N*-acylhomoserine lactone autoinducer. Proc Natl Acad Sci USA 1994; 91(11): 4639-43.
[http://dx.doi.org/10.1073/pnas.91.11.4639] [PMID: 8197112]

[26]   Ahlgren NA, Harwood CS, Schaefer AL, Giraud E, Greenberg EP. Aryl-homoserine lactone quorum sensing in stem-nodulating photosynthetic bradyrhizobia. Proc Natl Acad Sci USA 2011; 108(17): 7183-8.
[http://dx.doi.org/10.1073/pnas.1103821108] [PMID: 21471459]

[27]   Wang LH, He Y, Gao Y, *et al.* A bacterial cell-cell communication signal with cross-kingdom structural analogues. Mol Microbiol 2004; 51(3): 903-12.
[http://dx.doi.org/10.1046/j.1365-2958.2003.03883.x] [PMID: 14731288]

[28]   Ramsay JP, Sullivan JT, Stuart GS, Lamont IL, Ronson CW. Excision and transfer of the *Mesorhizobium loti* R7A symbiosis island requires an integrase IntS, a novel recombination directionality factor RdfS, and a putative relaxase RlxS. Mol Microbiol 2006; 62(3): 723-34.
[http://dx.doi.org/10.1111/j.1365-2958.2006.05396.x] [PMID: 17076666]

[29]   Déziel E, Lépine F, Milot S, *et al.* Analysis of *Pseudomonas aeruginosa* 4-hydroxy-2-alkylquinolines (HAQs) reveals a role for 4-hydroxy-2-heptylquinoline in cell-to-cell communication. Proc Natl Acad Sci USA 2004; 101(5): 1339-44.
[http://dx.doi.org/10.1073/pnas.0307694100] [PMID: 14739337]

[30]   Daniels R, De Vos DE, Desair J, *et al.* The cin quorum sensing locus of *Rhizobium etli* CNPAF512 affects growth and symbiotic nitrogen fixation. J Biol Chem 2002; 277(1): 462-8.
[http://dx.doi.org/10.1074/jbc.M106655200] [PMID: 11677232]

[31]   Sturgill G, Rather PN. Evidence that putrescine acts as an extracellular signal required for swarming in *Proteus mirabilis*. Mol Microbiol 2004; 51(2): 437-46.
[http://dx.doi.org/10.1046/j.1365-2958.2003.03835.x] [PMID: 14756784]

[32]   Marketon MM, González JE. Identification of two quorum-sensing systems in *Sinorhizobium meliloti*.

J Bacteriol 2002; 184(13): 3466-75.
[http://dx.doi.org/10.1128/JB.184.13.3466-3475.2002] [PMID: 12057940]

[33]   Dixon R, Kahn D. Genetic regulation of biological nitrogen fixation. Nat Rev Microbiol 2004; 2(8): 621-31.
[http://dx.doi.org/10.1038/nrmicro954] [PMID: 15263897]

[34]   Janssen PH. Identifying the dominant soil bacterial taxa in libraries of 16S rRNA and 16S rRNA genes. Appl Environ Microbiol 2006; 72(3): 1719-28.
[http://dx.doi.org/10.1128/AEM.72.3.1719-1728.2006] [PMID: 16517615]

[35]   An JH, Goo E, Kim H, Seo YS, Hwang I. Bacterial quorum sensing and metabolic slowing in a cooperative population. Proc Natl Acad Sci USA 2014; 111(41): 14912-7.
[http://dx.doi.org/10.1073/pnas.1412431111] [PMID: 25267613]

[36]   Sperandio V, Torres AG, Kaper JB. Quorum sensing *Escherichia coli* regulators B and C (QseBC): a novel two-component regulatory system involved in the regulation of flagella and motility by quorum sensing in *E. coli*. Mol Microbiol 2002; 43(3): 809-21.
[http://dx.doi.org/10.1046/j.1365-2958.2002.02803.x] [PMID: 11929534]

[37]   Chu W, Jiang Y, Yongwang L, Zhu W. Role of the quorum-sensing system in biofilm formation and virulence of *Aeromonas hydrophila*. Afr J Microbiol Res 2011; 5(32): 5819-25.

[38]   Rocha J, Flores V, Cabrera R, *et al.* Evolution and some functions of the NprR–NprRB quorum-sensing system in the *Bacillus cereus* group. Appl Microbiol Biotechnol 2012; 94(4): 1069-78.
[http://dx.doi.org/10.1007/s00253-011-3775-4] [PMID: 22159892]

[39]   Fuqua WC, Winans SC, Greenberg EP. Quorum sensing in bacteria: the LuxR-LuxI family of cell density-responsive transcriptional regulators. J Bacteriol 1994; 176(2): 269-75.
[http://dx.doi.org/10.1128/jb.176.2.269-275.1994] [PMID: 8288518]

[40]   Miller MB, Bassler BL. Quorum sensing in bacteria. Annu Rev Microbiol 2001; 55(1): 165-99.
[http://dx.doi.org/10.1146/annurev.micro.55.1.165] [PMID: 11544353]

[41]   Ng WL, Bassler BL. Bacterial quorum-sensing network architectures. Annu Rev Genet 2009; 43(1): 197-222.
[http://dx.doi.org/10.1146/annurev-genet-102108-134304] [PMID: 19686078]

[42]   Kleerebezem M, Quadri LEN, Kuipers OP, De Vos WM. Quorum sensing by peptide pheromones and two-component signal-transduction systems in Gram-positive bacteria. Mol Microbiol 1997; 24(5): 895-904.
[http://dx.doi.org/10.1046/j.1365-2958.1997.4251782.x] [PMID: 9219998]

[43]   Pathak DT, Wei X, Dey A, Wall D. Molecular recognition by a polymorphic cell surface receptor governs cooperative behaviors in bacteria. PLoS Genet 2013; 9(11): e1003891.
[http://dx.doi.org/10.1371/journal.pgen.1003891] [PMID: 24244178]

[44]   Sah GP, Wall D. Kin recognition and outer membrane exchange (OME) in myxobacteria. Curr Opin Microbiol 2020; 56: 81-8.
[http://dx.doi.org/10.1016/j.mib.2020.07.003] [PMID: 32828979]

[45]   Cao P, Wall D. Self-identity reprogrammed by a single residue switch in a cell surface receptor of a social bacterium. Proc Natl Acad Sci USA 2017; 114(14): 3732-7.
[http://dx.doi.org/10.1073/pnas.1700315114] [PMID: 28320967]

[46]   He X, Chang W, Pierce DL, Seib LO, Wagner J, Fuqua C. Quorum sensing in *Rhizobium* sp. strain NGR234 regulates conjugal transfer (tra) gene expression and influences growth rate. J Bacteriol 2003; 185(3): 809-22.
[http://dx.doi.org/10.1128/JB.185.3.809-822.2003] [PMID: 12533456]

[47]   Liu F, Zhao Q, Jia Z, *et al.* N-3-oxo-octanoyl-homoserine lactone-mediated priming of resistance to *Pseudomonas syringae* requires the salicylic acid signaling pathway in *Arabidopsis thaliana.* BMC Plant Biol 2020; 20(1): 38.

[http://dx.doi.org/10.1186/s12870-019-2228-6] [PMID: 31992205]

[48]   Fuqua C, Greenberg EP. Listening in on bacteria: acyl-homoserine lactone signalling. Nat Rev Mol Cell Biol 2002; 3(9): 685-95.
       [http://dx.doi.org/10.1038/nrm907] [PMID: 12209128]

[49]   Jitacksorn S, Sadowsky MJ. Nodulation gene regulation and quorum sensing control density-dependent suppression and restriction of nodulation in the Bradyrhizobium japonicum-soybean symbiosis. Appl Environ Microbiol 2008; 74(12): 3749-56.
       [http://dx.doi.org/10.1128/AEM.02939-07] [PMID: 18441104]

[50]   Pongsilp N, Triplett EW, Sadowsky MJ. Detection of homoserine lactone-like quorum sensing molecules in *bradyrhizobium* strains. Curr Microbiol 2005; 51(4): 250-4.
       [http://dx.doi.org/10.1007/s00284-005-4550-5] [PMID: 16132457]

[51]   Churchill MEA, Sibhatu HM, Uhlson CL. Defining the structure and function of acyl-homoserine lactone autoinducers. In: Rumbaugh K, Ed. Quorum Sensing Methods in Molecular Biology (Methods and Protocols). Humana Press 2011; Vol. 692.
       [http://dx.doi.org/10.1007/978-1-60761-971-0_12]

[52]   Gao Y, Zhong Z, Sun K, Wang H, Zhu J. The quorum-sensing system in a plant bacterium *Mesorhizobium huakuii* affects growth rate and symbiotic nodulation. Plant Soil 2006; 286(1-2): 53-60.
       [http://dx.doi.org/10.1007/s11104-006-9025-3]

[53]   Papenfort K, Bassler BL. Quorum sensing signal–response systems in Gram-negative bacteria. Nat Rev Microbiol 2016; 14(9): 576-88.
       [http://dx.doi.org/10.1038/nrmicro.2016.89] [PMID: 27510864]

[54]   Cai W, Ou F, Staehelin C, Dai W. BRADYRHIZOBIUM sp. strain ORS278 promotes rice growth and its quorum sensing system is required for optimal root colonization. Environ Microbiol Rep 2020; 12(6): 656-66.
       [http://dx.doi.org/10.1111/1758-2229.12885] [PMID: 32929871]

[55]   Ramsay JP, Sullivan JT, Jambari N, *et al.* A LuxRI-family regulatory system controls excision and transfer of the *Mesorhizobium loti* strain R7A symbiosis island by activating expression of two conserved hypothetical genes. Mol Microbiol 2009; 73(6): 1141-55.
       [http://dx.doi.org/10.1111/j.1365-2958.2009.06843.x] [PMID: 19682258]

[56]   Yan L, Allen MS, Simpson ML, Sayler GS, Cox CD. Direct quantification of *N*-(3-oxo-hexanoy-)-l-homoserine lactone in culture supernatant using a whole-cell bioreporter. J Microbiol Methods 2007; 68(1): 40-5.
       [http://dx.doi.org/10.1016/j.mimet.2006.06.002] [PMID: 16916554]

[57]   Daniels R, Reynaert S, Hoekstra H, *et al.* Quorum signal molecules as biosurfactants affecting swarming in *Rhizobium etli*. Proc Natl Acad Sci USA 2006; 103(40): 14965-70.
       [http://dx.doi.org/10.1073/pnas.0511037103] [PMID: 16990436]

[58]   Suárez-Moreno ZR, Caballero-Mellado J, Venturi V. The new group of non-pathogenic plant-associated nitrogen-fixing *Burkholderia* spp. shares a conserved quorum-sensing system, which is tightly regulated by the RsaL repressor. Microbiology (Reading) 2008; 154(7): 2048-59.
       [http://dx.doi.org/10.1099/mic.0.2008/017780-0] [PMID: 18599833]

[59]   Suárez-Moreno ZR, Devescovi G, Myers M, *et al.* Commonalities and differences in regulation of *N*-acyl homoserine lactone quorum sensing in the beneficial plant-associated *burkholderia* species cluster. Appl Environ Microbiol 2010; 76(13): 4302-17.
       [http://dx.doi.org/10.1128/AEM.03086-09] [PMID: 20435760]

[60]   Loh J, Carlson RW, York WS, Stacey G. Bradyoxetin, a unique chemical signal involved in symbiotic gene regulation. Proc Natl Acad Sci USA 2002; 99(22): 14446-51.
       [http://dx.doi.org/10.1073/pnas.222336799] [PMID: 12393811]

[61]   Loh J, Stacey G. Nodulation gene regulation in *Bradyrhizobium japonicum*: a unique integration of global regulatory circuits. Appl Environ Microbiol 2003; 69(1): 10-7.
[http://dx.doi.org/10.1128/AEM.69.1.10-17.2003] [PMID: 12513971]

[62]   Lazazzera BA, Lazazzera BA. The extracellular PHR peptide-Rap phosphatase signaling circuit of *bacillus subtilis*. Front Biosci 2003; 8(4): 913.
[http://dx.doi.org/10.2741/913] [PMID: 12456319]

[63]   Novick RP. Regulation of pathogenicity in Staphylococcus aureus by a peptide-based density-sensing system Cell-cell signaling in bacteria. DC: American Society for Microbiology Washington 1999; pp. 129-46.

[64]   Schauder S, Shokat K, Surette MG, Bassler BL. The LuxS family of bacterial autoinducers: biosynthesis of a novel quorum-sensing signal molecule. Mol Microbiol 2001; 41(2): 463-76.
[http://dx.doi.org/10.1046/j.1365-2958.2001.02532.x] [PMID: 11489131]

[65]   Sun J, Daniel R, Wagner-Döbler I, Zeng AP. Is autoinducer-2 a universal signal for interspecies communication: a comparative genomic and phylogenetic analysis of the synthesis and signal transduction pathways. BMC Evol Biol 2004; 4(1): 36.
[http://dx.doi.org/10.1186/1471-2148-4-36] [PMID: 15456522]

[66]   Deng Y, Wu J, Eberl L, Zhang LH. Structural and functional characterization of diffusible signal factor family quorum-sensing signals produced by members of the *Burkholderia cepacia* complex. Appl Environ Microbiol 2010; 76(14): 4675-83.
[http://dx.doi.org/10.1128/AEM.00480-10] [PMID: 20511428]

[67]   Bredenbruch F, Nimtz M, Wray V, Morr M, Müller R, Häussler S. Biosynthetic pathway of *Pseudomonas aeruginosa* 4-hydroxy-2-alkylquinolines. J Bacteriol 2005; 187(11): 3630-5.
[http://dx.doi.org/10.1128/JB.187.11.3630-3635.2005] [PMID: 15901684]

[68]   Flavier AB, Clough SJ, Schell MA, Denny TP. Identification of 3-hydroxypalmitic acid methyl ester as a novel autoregulator controlling virulence in *Ralstonia solanacearum*. Mol Microbiol 1997; 26(2): 251-9.
[http://dx.doi.org/10.1046/j.1365-2958.1997.5661945.x] [PMID: 9383151]

[69]   Lithgow JK, Wilkinson A, Hardman A, *et al.* The regulatory locus cinRI in *Rhizobium leguminosarum* controls a network of quorum-sensing loci. Mol Microbiol 2000; 37(1): 81-97.
[http://dx.doi.org/10.1046/j.1365-2958.2000.01960.x] [PMID: 10931307]

[70]   Wilkinson A, Danino V, Wisniewski-Dyé F, Lithgow JK, Downie JA. *N*-acyl-homoserine lactone inhibition of rhizobial growth is mediated by two quorum-sensing genes that regulate plasmid transfer. J Bacteriol 2002; 184(16): 4510-9.
[http://dx.doi.org/10.1128/JB.184.16.4510-4519.2002] [PMID: 12142421]

[71]   Kaiser D. Signaling in myxobacteria. Annu Rev Microbiol 2004; 58(1): 75-98.
[http://dx.doi.org/10.1146/annurev.micro.58.030603.123620] [PMID: 15487930]

[72]   Bassler BL. Small Talk. Cell 2002; 109(4): 421-4.
[http://dx.doi.org/10.1016/S0092-8674(02)00749-3] [PMID: 12086599]

[73]   Christensen AB, Riedel K, Eberl L, *et al.* Quorum-sensing-directed protein expression in *Serratia proteamaculans* B5a. Microbiology (Reading) 2003; 149(2): 471-83.
[http://dx.doi.org/10.1099/mic.0.25575-0] [PMID: 12624209]

[74]   Watson WT, Minogue TD, Val DL, von Bodman SB, Churchill MEA. Structural basis and specificity of acyl-homoserine lactone signal production in bacterial quorum sensing. Mol Cell 2002; 9(3): 685-94.
[http://dx.doi.org/10.1016/S1097-2765(02)00480-X] [PMID: 11931774]

[75]   Vincent D, Rafiqi M, Job D. The multiple facets of plant–fungal interactions revealed through plant and fungal secretomics. Front Plant Sci 2020; 10: 1626.
[http://dx.doi.org/10.3389/fpls.2019.01626] [PMID: 31969889]

[76]   Szabó Z, Tönnis M, Kessler H, Feldbrügge M. Structure-function analysis of lipopeptide pheromones from the plant pathogen *Ustilago maydis*. Mol Genet Genomics 2002; 268(3): 362-70.
[http://dx.doi.org/10.1007/s00438-002-0756-4] [PMID: 12436258]

[77]   Liang Y, Tóth K, Cao Y, Tanaka K, Espinoza C, Stacey G. Lipochitooligosaccharide recognition: an ancient story. New Phytol 2014; 204(2): 289-96.
[http://dx.doi.org/10.1111/nph.12898] [PMID: 25453133]

[78]   Tsitsigiannis DI, Keller NP. Oxylipins as developmental and host–fungal communication signals. Trends Microbiol 2007; 15(3): 109-18.
[http://dx.doi.org/10.1016/j.tim.2007.01.005] [PMID: 17276068]

[79]   Kanchiswamy CN, Malnoy M, Maffei ME. Chemical diversity of microbial volatiles and their potential for plant growth and productivity. Front Plant Sci 2015; 6: 151.
[http://dx.doi.org/10.3389/fpls.2015.00151] [PMID: 25821453]

[80]   Gao X, Kolomiets MV. Host-derived lipids and oxylipins are crucial signals in modulating mycotoxin production by fungi. Toxin Rev 2009; 28(2-3): 79-88.
[http://dx.doi.org/10.1080/15569540802420584]

[81]   Kaiser R. Flowers and fungi use scents to mimic each other. Science 2006; 311(5762): 806-7.
[http://dx.doi.org/10.1126/science.1119499] [PMID: 16469916]

[82]   Schmidt R, Etalo DW, de Jager V, *et al.* Microbial small talk: volatiles in fungal–bacterial interactions. Front Microbiol 2016; 6: 1495.
[PMID: 26779150]

[83]   Li N, Alfiky A, Vaughan MM, Kang S. Stop and smell the fungi: Fungal volatile metabolites are overlooked signals involved in fungal interaction with plants. Fungal Biol Rev 2016; 30(3): 134-44.
[http://dx.doi.org/10.1016/j.fbr.2016.06.004]

[84]   Leach JE, Triplett LR, Argueso CT, Trivedi P. Communication in the Phytobiome. Cell 2017; 169(4): 587-96.
[http://dx.doi.org/10.1016/j.cell.2017.04.025] [PMID: 28475891]

[85]   Parniske M. Arbuscular mycorrhiza: the mother of plant root endosymbioses. Nat Rev Microbiol 2008; 6(10): 763-75.
[http://dx.doi.org/10.1038/nrmicro1987] [PMID: 18794914]

[86]   Navazio L, Mariani P. Calcium opens the dialogue between plants and arbuscular mycorrhizal fungi. Plant Signal Behav 2008; 3(4): 229-30.
[http://dx.doi.org/10.4161/psb.3.4.5093] [PMID: 19704636]

[87]   Nanjundappa A, Bagyaraj DJ, Saxena AK, Kumar M, Chakdar H. Interaction between arbuscular mycorrhizal fungi and *Bacillus* spp. in soil enhancing growth of crop plants. Fungal Biol Biotechnol 2019; 6(1): 23.
[http://dx.doi.org/10.1186/s40694-019-0086-5] [PMID: 31798924]

[88]   Nakkeeran S, Rajamanickam S, Vanthana M, Renukadevi P, Muthamilan M. Harnessing the perception of Trichoderma signal molecules in rhizosphere to improve soil health and plant health. 2020.
[http://dx.doi.org/10.1007/978-981-15-3321-1_4]

[89]   Brotman Y, Briff E, Viterbo A, Chet I. Role of swollenin, an expansin-like protein from Trichoderma, in plant root colonization. Plant Physiol 2008; 147(2): 779-89.
[http://dx.doi.org/10.1104/pp.108.116293] [PMID: 18400936]

[90]   Mukherjee M, Mukherjee PK, Horwitz BA, Zachow C, Berg G, Zeilinger S. Trichoderma-plan--pathogen interactions: advances in genetics of biological control. Indian J Microbiol 2012; 52(4): 522-9.
[http://dx.doi.org/10.1007/s12088-012-0308-5] [PMID: 24293705]

[91]   Horinouchi S. A microbial hormone, A-factor, as a master switch for morphological differentiation and secondary metabolism in *Streptomyces griseus*. Front Biosci 2002; 7: d2045-57.
        [PMID: 12165483]

[92]   Chen X, Lu Y, Fan Y, Shen Y. Production of validamycins 2017.
        [http://dx.doi.org/10.1016/B978-0-08-100999-4.00002-2]

[93]   Lanfranco L, Fiorilli V, Venice F, Bonfante P. Strigolactones cross the kingdoms: plants, fungi, and bacteria in the arbuscular mycorrhizal symbiosis. J Exp Bot 2018; 69(9): 2175-88.
        [http://dx.doi.org/10.1093/jxb/erx432] [PMID: 29309622]

[94]   Ehlers BK, Berg MP, Staudt M, *et al.* Plant secondary compounds in soil and their role in below-ground species interactions. Trends Ecol Evol 2020; 35(8): 716-30.
        [http://dx.doi.org/10.1016/j.tree.2020.04.001] [PMID: 32414604]

[95]   Aniszewski T. Alkaloids: chemistry, biology, ecology, and applications. Elsevier 2015.

[96]   de Rijke E, Out P, Niessen WMA, Ariese F, Gooijer C, Brinkman UAT. Analytical separation and detection methods for flavonoids. J Chromatogr A 2006; 1112(1-2): 31-63.
        [http://dx.doi.org/10.1016/j.chroma.2006.01.019] [PMID: 16480997]

[97]   Weston LA, Mathesius U. Flavonoids: their structure, biosynthesis and role in the rhizosphere, including allelopathy. J Chem Ecol 2013; 39(2): 283-97.
        [http://dx.doi.org/10.1007/s10886-013-0248-5] [PMID: 23397456]

[98]   Heath MC. Hypersensitive response-related death. Plant Mol Biol 2000; 44(3): 321-34.
        [http://dx.doi.org/10.1023/A:1026592509060] [PMID: 11199391]

[99]   Wezel AP, Opperhuizen A. Narcosis due to environmental pollutants in aquatic organisms: residue-based toxicity, mechanisms, and membrane burdens. Crit Rev Toxicol 1995; 25(3): 255-79.
        [http://dx.doi.org/10.3109/10408449509089890] [PMID: 7576154]

[100]   Glasius M, Goldstein AH. Recent discoveries and future challenges in atmospheric organic chemistry. Environ Sci Technol 2016; 50(6): 2754-64.
        [http://dx.doi.org/10.1021/acs.est.5b05105] [PMID: 26862779]

[101]   Halkier BA, Gershenzon J. Biology and biochemistry of glucosinolates. Annu Rev Plant Biol 2006; 57(1): 303-33.
        [http://dx.doi.org/10.1146/annurev.arplant.57.032905.105228] [PMID: 16669764]

[102]   Jacoby RP, Kopriva S. Metabolic niches in the rhizosphere microbiome: new tools and approaches to analyse metabolic mechanisms of plant–microbe nutrient exchange. J Exp Bot 2019; 70(4): 1087-94.
        [http://dx.doi.org/10.1093/jxb/ery438] [PMID: 30576534]

[103]   van der Heijden MGA, Bardgett RD, van Straalen NM. The unseen majority: soil microbes as drivers of plant diversity and productivity in terrestrial ecosystems. Ecol Lett 2008; 11(3): 296-310.
        [http://dx.doi.org/10.1111/j.1461-0248.2007.01139.x] [PMID: 18047587]

[104]   Mendes R, Garbeva P, Raaijmakers JM. The rhizosphere microbiome: significance of plant beneficial, plant pathogenic, and human pathogenic microorganisms. FEMS Microbiol Rev 2013; 37(5): 634-63.
        [http://dx.doi.org/10.1111/1574-6976.12028] [PMID: 23790204]

[105]   Verbon EH, Liberman LM. Beneficial microbes affect endogenous mechanisms controlling root development. Trends Plant Sci 2016; 21(3): 218-29.
        [http://dx.doi.org/10.1016/j.tplants.2016.01.013] [PMID: 26875056]

[106]   Bonkowski M. Protozoa and plant growth: the microbial loop in soil revisited. New Phytol 2004; 162(3): 617-31.
        [http://dx.doi.org/10.1111/j.1469-8137.2004.01066.x] [PMID: 33873756]

[107]   Richardson AE, Barea JM, McNeill AM, Prigent-Combaret C. Acquisition of phosphorus and nitrogen in the rhizosphere and plant growth promotion by microorganisms. Plant Soil 2009; 321(1-2): 305-39.
        [http://dx.doi.org/10.1007/s11104-009-9895-2]

[108] Schimel JP, Bennett J. Nitrogen mineralization: challenges of a changing paradigm. Ecology 2004; 85(3): 591-602.
[http://dx.doi.org/10.1890/03-8002]

[109] Shiraishi A, Matsushita N, Hougetsu T. Nodulation in black locust by the Gammaproteobacteria *Pseudomonas* sp. and the Betaproteobacteria *Burkholderia* sp. Syst Appl Microbiol 2010; 33(5): 269-74.
[http://dx.doi.org/10.1016/j.syapm.2010.04.005] [PMID: 20542651]

[110] Masson-Boivin C, Giraud E, Perret X, Batut J. Establishing nitrogen-fixing symbiosis with legumes: how many rhizobium recipes? Trends Microbiol 2009; 17(10): 458-66.
[http://dx.doi.org/10.1016/j.tim.2009.07.004] [PMID: 19766492]

[111] Oldroyd GED, Murray JD, Poole PS, Downie JA. The rules of engagement in the legume-rhizobial symbiosis. Annu Rev Genet 2011; 45(1): 119-44.
[http://dx.doi.org/10.1146/annurev-genet-110410-132549] [PMID: 21838550]

[112] Santi C, Bogusz D, Franche C. Biological nitrogen fixation in non-legume plants. Ann Bot (Lond) 2013; 111(5): 743-67.
[http://dx.doi.org/10.1093/aob/mct048] [PMID: 23478942]

[113] Brundrett M. Diversity and classification of mycorrhizal associations. Biol Rev Camb Philos Soc 2004; 79(3): 473-95.
[http://dx.doi.org/10.1017/S1464793103006316] [PMID: 15366760]

[114] Roth-Bejerano N, Navarro-Rodenas A, Gutierrez A. Types of mycorrhizal association. In: Kagan-Zur V, Roth-Bejerano N, Sitrit Y, Morte A, Eds. Desert Truffles Soil Biology. Berlin, Heidelberg: Springer 2014; Vol. 38.
[http://dx.doi.org/10.1007/978-3-642-40096-4_5]

[115] Barman J, Samanta A, Saha B, Datta S. Mycorrhiza the oldest association between plant and fungi. Reson J Sci Educ 2016; 21(12): 1093-04.

[116] Gadkar V, David-Schwartz R, Kunik T, Kapulnik Y. Arbuscular mycorrhizal fungal colonization. Factors involved in host recognition. Plant Physiol 2001; 127(4): 1493-9.
[http://dx.doi.org/10.1104/pp.010783] [PMID: 11743093]

[117] Morkunas I, Woźniak A, Mai V, Rucińska-Sobkowiak R, Jeandet P. The role of heavy metals in plant response to biotic stress. Molecules 2018; 23(9): 2320.
[http://dx.doi.org/10.3390/molecules23092320] [PMID: 30208652]

[118] Newsham KK. A meta-analysis of plant responses to dark septate root endophytes. New Phytol 2011; 190(3): 783-93.
[http://dx.doi.org/10.1111/j.1469-8137.2010.03611.x] [PMID: 21244432]

[119] Lin LC, Ye YS, Lin WR. Characteristics of root-cultivable endophytic fungi from *Rhododendron ovatum* Planch. Braz J Microbiol 2019; 50(1): 185-93.
[http://dx.doi.org/10.1007/s42770-018-0011-8] [PMID: 30637639]

[120] Wang W, Vinocur B, Altman A. Plant responses to drought, salinity and extreme temperatures: towards genetic engineering for stress tolerance. Planta 2003; 218(1): 1-14.
[http://dx.doi.org/10.1007/s00425-003-1105-5] [PMID: 14513379]

[121] Waqas MA, Kaya C, Riaz A, *et al.* Potential mechanisms of abiotic stress tolerance in crop plants induced by thiourea. Front Plant Sci 2019; 10: 1336.
[http://dx.doi.org/10.3389/fpls.2019.01336] [PMID: 31736993]

[122] Van Oosten MJ, Pepe O, De Pascale S, Silletti S, Maggio A. The role of biostimulants and bioeffectors as alleviators of abiotic stress in crop plants. Chem Biol Technol Agric 2017; 4(1): 5.
[http://dx.doi.org/10.1186/s40538-017-0089-5]

[123] Meena KK, Sorty AM, Bitla UM, *et al.* Abiotic stress responses and microbe-mediated mitigation in

plants: the omics strategies. Front Plant Sci 2017; 8: 172.
[http://dx.doi.org/10.3389/fpls.2017.00172] [PMID: 28232845]

[124] Zhu JK. Abiotic stress signaling and response in plants. Cell 2016; 167(2): 313-24.
[http://dx.doi.org/10.1016/j.cell.2016.08.029] [PMID: 27716505]

[125] Choudhary DK, Kasotia A, Jain S, *et al.* bacterial-mediated tolerance and resistance to plants under
abiotic and biotic stresses. J Plant Growth Regul 2016; 35(1): 276-300.
[http://dx.doi.org/10.1007/s00344-015-9521-x]

[126] Suyal DC, Yadav A, Shouche Y, Goel R. Differential proteomics in response to low temperature
diazotrophy of Himalayan psychrophilic nitrogen fixing Pseudomonas migulae S10724 strain. Curr
Microbiol 2014; 68(4): 543-50.
[http://dx.doi.org/10.1007/s00284-013-0508-1] [PMID: 24362552]

[127] Suyal DC, Kumar S, Joshi D, Soni R, Goel R. Quantitative proteomics of psychotrophic diazotroph in
response to nitrogen deficiency and cold stress. J Proteomics 2018; 187: 235-42.
[http://dx.doi.org/10.1016/j.jprot.2018.08.005] [PMID: 30092381]

[128] Latha PK, Soni R, Khan M, Marla SS, Goel R. Exploration of Csp genes from temperate and glacier
soils of the Indian Himalayas and *in silico* analysis of encoding proteins. Curr Microbiol 2009; 58(4):
343-8.
[http://dx.doi.org/10.1007/s00284-008-9344-0] [PMID: 19159976]

[129] Lindquist JA, Mertens PR. Cold shock proteins: from cellular mechanisms to pathophysiology and
disease. Cell Commun Signal 2018; 16(1): 63.
[http://dx.doi.org/10.1186/s12964-018-0274-6] [PMID: 30257675]

[130] Newkirk K, Feng W, Jiang W, *et al.* Solution NMR structure of the major cold shock protein (CspA)
from *Escherichia coli*: identification of a binding epitope for DNA. Proc Natl Acad Sci USA 1994;
91(11): 5114-8.
[http://dx.doi.org/10.1073/pnas.91.11.5114] [PMID: 7515185]

[131] Mueller U, Perl D, Schmid FX, Heinemann U. Thermal stability and atomic-resolution crystal
structure of the *Bacillus caldolyticus* cold shock protein. J Mol Biol 2000; 297(4): 975-88.
[http://dx.doi.org/10.1006/jmbi.2000.3602] [PMID: 10736231]

[132] Kremer W, Schuler B, Harrieder S, *et al.* Solution NMR structure of the cold-shock protein from the
hyperthermophilic bacterium *Thermotoga maritima*. Eur J Biochem 2001; 268(9): 2527-39.
[http://dx.doi.org/10.1046/j.1432-1327.2001.02127.x] [PMID: 11322871]

[133] Berg G, Köberl M, Rybakova D, Müller H, Grosch R, Smalla K. Plant microbial diversity is suggested
as the key to future biocontrol and health trends. FEMS Microbiol Ecol 2017; 93(5)
[http://dx.doi.org/10.1093/femsec/fix050] [PMID: 28430944]

[134] Vassilev N, Vassileva M, Lopez A, *et al.* Unexploited potential of some biotechnological techniques
for biofertilizer production and formulation. Appl Microbiol Biotechnol 2015; 99(12): 4983-96.
[http://dx.doi.org/10.1007/s00253-015-6656-4] [PMID: 25957155]

[135] Ortíz-Castro R, Contreras-Cornejo HA, Macías-Rodríguez L, López-Bucio J. The role of microbial
signals in plant growth and development. Plant Signal Behav 2009; 4(8): 701-12.
[http://dx.doi.org/10.4161/psb.4.8.9047] [PMID: 19820333]

[136] Confortin TC, Spannemberg SS, Todero I, *et al.* microbial enzymes as control agents of diseases and
pests in organic agriculture. In: Gupta VK, Pandey A, Eds. New and Future Developments in
Microbial Biotechnology and Bioengineering. Elsevier 2019; pp. 321-32.
[http://dx.doi.org/10.1016/B978-0-444-63504-4.00021-9]

[137] Chatterton S, Punja ZK. Chitinase and β-1,3-glucanase enzyme production by the mycoparasite
Clonostachys rosea f. *catenulata* against fungal plant pathogens. Can J Microbiol 2009; 55(4): 356-67.
[http://dx.doi.org/10.1139/W08-156] [PMID: 19396235]

[138] Onofre-Lemus J, Hernández-Lucas I, Girard L, Caballero-Mellado J. ACC (1-aminocyclopropane-

1-carboxylate) deaminase activity, a widespread trait in *Burkholderia* species, and its growth-promoting effect on tomato plants. Appl Environ Microbiol 2009; 75(20): 6581-90.
[http://dx.doi.org/10.1128/AEM.01240-09] [PMID: 19700546]

[139]   Rijavec T, Lapanje A. Hydrogen cyanide in the rhizosphere: not suppressing plant pathogens, but rather regulating availability of phosphate. Front Microbiol 2016; 7: 1785.
[http://dx.doi.org/10.3389/fmicb.2016.01785] [PMID: 27917154]

[140]   Olanrewaju OS, Glick BR, Babalola OO. Mechanisms of action of plant growth promoting bacteria. World J Microbiol Biotechnol 2017; 33(11): 197.
[http://dx.doi.org/10.1007/s11274-017-2364-9] [PMID: 28986676]

[141]   Yu JM, Wang D, Pierson LS III, Pierson EA. Effect of producing different phenazines on bacterial fitness and biological control in *Pseudomonas chlororaphis* 30-84. Plant Pathol J 2018; 34(1): 44-58.
[http://dx.doi.org/10.5423/PPJ.FT.12.2017.0277] [PMID: 29422787]

[142]   Biessy A, Filion M. Phenazines in plant-beneficial *Pseudomonas* spp.: biosynthesis, regulation, function and genomics. Environ Microbiol 2018; 20(11): 3905-17.
[http://dx.doi.org/10.1111/1462-2920.14395] [PMID: 30159978]

[143]   Tripathi RK, Gottlieb D. Mechanism of action of the antifungal antibiotic pyrrolnitrin. J Bacteriol 1969; 100(1): 310-8.
[http://dx.doi.org/10.1128/jb.100.1.310-318.1969] [PMID: 4310080]

[144]   Pawar S, Chaudhari A, Prabha R, Shukla R, Singh DP. Microbial pyrrolnitrin: natural metabolite with immense practical utility. Biomolecules 2019; 9(9): 443.
[http://dx.doi.org/10.3390/biom9090443] [PMID: 31484394]

[145]   Almario J, Bruto M, Vacheron J, Prigent-Combaret C, Moënne-Loccoz Y, Muller D. Distribution of 2,4-diacetylphloroglucinol biosynthetic genes among the *Pseudomonas* spp. reveals unexpected polyphyletism. Front Microbiol 2017; 8: 1218.
[http://dx.doi.org/10.3389/fmicb.2017.01218] [PMID: 28713346]

[146]   Keswani C, Singh HB, García-Estrada C, *et al.* Antimicrobial secondary metabolites from agriculturally important bacteria as next-generation pesticides. Appl Microbiol Biotechnol 2020; 104(3): 1013-34.
[http://dx.doi.org/10.1007/s00253-019-10300-8] [PMID: 31858191]

[147]   Nielsen TH, Christophersen C, Anthoni U, Sørensen J. Viscosinamide, a new cyclic depsipeptide with surfactant and antifungal properties produced by *Pseudomonas fluorescens* DR54. J Appl Microbiol 1999; 87(1): 80-90.
[http://dx.doi.org/10.1046/j.1365-2672.1999.00798.x] [PMID: 10432590]

[148]   Ho YN, Lee HJ, Hsieh CT, Peng CC, Yang YL. Chemistry and biology of salicyl-capped siderophores. 2018.
[http://dx.doi.org/10.1016/B978-0-444-64179-3.00013-X]

[149]   Gurusinghe S, Brooks TL, Barrow RA, *et al.* Technologies for the selection, culture and metabolic profiling of unique rhizosphere microorganisms for natural product discovery. Molecules 2019; 24(10): 1955.
[http://dx.doi.org/10.3390/molecules24101955] [PMID: 31117282]

[150]   Schöner TA, Fuchs SW, Reinhold-Hurek B, Bode HB. Identification and biosynthesis of a novel xanthomonadin-dialkylresorcinol-hybrid from *Azoarcus* sp. BH72. PLoS One 2014; 9(3): e90922.
[http://dx.doi.org/10.1371/journal.pone.0090922] [PMID: 24618669]

[151]   Gross H, Loper JE. Genomics of secondary metabolite production by *Pseudomonas* spp. Nat Prod Rep 2009; 26(11): 1408-46.
[http://dx.doi.org/10.1039/b817075b] [PMID: 19844639]

[152]   Loper JE, Henkels MD, Shaffer BT, Valeriote FA, Gross H. Isolation and identification of rhizoxin analogs from *Pseudomonas fluorescens* Pf-5 by using a genomic mining strategy. Appl Environ

Microbiol 2008; 74(10): 3085-93.
[http://dx.doi.org/10.1128/AEM.02848-07] [PMID: 18344330]

[153] El-Sayed AK, Hothersall J, Thomas CM. Quorum-sensing-dependent regulation of biosynthesis of the polyketide antibiotic mupirocin in *Pseudomonas fluorescens* NCIMB 10586 The GenBank accession numbers for the sequences determined in this work are AF318063 (mupA), AF318064 (mupR) and AF318065 (mupI). Microbiology (Reading) 2001; 147(8): 2127-39.
[http://dx.doi.org/10.1099/00221287-147-8-2127] [PMID: 11495990]

[154] Capobianco JO, Doran CC, Goldman RC. Mechanism of mupirocin transport into sensitive and resistant bacteria. Antimicrob Agents Chemother 1989; 33(2): 156-63.
[http://dx.doi.org/10.1128/AAC.33.2.156] [PMID: 2497702]

[155] Okrent RA, Trippe KM, Maselko M, Manning V. Functional analysis of a biosynthetic cluster essential for production of 4-formylaminooxyvinylglycine, a germination-arrest factor from *Pseudomonas fluorescens* WH6. bioRxiv 2006; 080572.
[http://dx.doi.org/10.1101/080572]

[156] Gross H, Stockwell VO, Henkels MD, Nowak-Thompson B, Loper JE, Gerwick WH. The genomisotopic approach: a systematic method to isolate products of orphan biosynthetic gene clusters. Chem Biol 2007; 14(1): 53-63.
[http://dx.doi.org/10.1016/j.chembiol.2006.11.007] [PMID: 17254952]

[157] Jang JY, Yang SY, Kim YC, *et al.* Identification of orfamide A as an insecticidal metabolite produced by *Pseudomonas protegens* F6. J Agric Food Chem 2013; 61(28): 6786-91.
[http://dx.doi.org/10.1021/jf401218w] [PMID: 23763636]

[158] Oni FE, Olorunleke OF, Höfte M. Phenazines and cyclic lipopeptides produced by Pseudomonas sp. CMR12a are involved in the biological control of Pythium myriotylum on cocoyam (Xanthosoma sagittifolium). Biol Control 2019; 129: 109-14.
[http://dx.doi.org/10.1016/j.biocontrol.2018.10.005]

[159] Ma Z, Zhang S, Liang J, Sun K, Hu J. Isolation and characterization of a new cyclic lipopeptide orfamide H from *Pseudomonas protegens* CHA0. J Antibiot (Tokyo) 2020; 73(3): 179-83.
[http://dx.doi.org/10.1038/s41429-019-0254-0] [PMID: 31666660]

[160] Simionato AS, Navarro MOP, de Jesus MLA, *et al.* The effect of phenazine-1-carboxylic acid on mycelial growth of *Botrytis cinerea* produced by *Pseudomonas aeruginosa* LV strain. Front Microbiol 2017; 8: 1102.
[http://dx.doi.org/10.3389/fmicb.2017.01102] [PMID: 28659907]

[161] Morrison CK, Arseneault T, Novinscak A, Filion M. Phenazine-1-carboxylic acid production by *Pseudomonas fluorescens* LBUM636 alters *Phytophthora infestans* growth and late blight development. Phytopathology 2017; 107(3): 273-9.
[http://dx.doi.org/10.1094/PHYTO-06-16-0247-R] [PMID: 27827009]

[162] Trippe K, McPhail K, Armstrong D, Azevedo M, Banowetz G. *Pseudomonas fluorescens* SBW25 produces furanomycin, a non-proteinogenic amino acid with selective antimicrobial properties. BMC Microbiol 2013; 13(1): 111.
[http://dx.doi.org/10.1186/1471-2180-13-111] [PMID: 23688329]

[163] Masschelein J, Jenner M, Challis GL. Antibiotics from Gram-negative bacteria: a comprehensive overview and selected biosynthetic highlights. Nat Prod Rep 2017; 34(7): 712-83.
[http://dx.doi.org/10.1039/C7NP00010C] [PMID: 28650032]

[164] Schmidt Y, van der Voort M, Crüsemann M, *et al.* Biosynthetic origin of the antibiotic cyclocarbamate brabantamide A (SB-253514) in plant-associated *Pseudomonas*. ChemBioChem 2014; 15(2): 259-66.
[http://dx.doi.org/10.1002/cbic.201300527] [PMID: 24436210]

[165] Pu Y, Lowe C, Sailer M, Vederas JC. Synthesis, stability and antimicrobial activity of (+)-obafluorin and related beta-lactone antibiotics. J Org Chem 1994; 59(13): 3642-55.
[http://dx.doi.org/10.1021/jo00092a025]

[166] Ye L, Cornelis P, Guillemyn K, Ballet S, Christophersen C, Hammerich O. Structure Revision of *N* - Mercapto-4-formylcarbostyril Produced by *Pseudomonas fluorescens* G308 to 2-(2-Hydroxyphenyl)thiazole-4-carbaldehyde [aeruginaldehyde]. Nat Prod Commun 2014; 9(6): 1934578X1400900.
[http://dx.doi.org/10.1177/1934578X1400900615] [PMID: 25115080]

[167] Santos Kron A, Zengerer V, Bieri M, *et al. Pseudomonas orientalis* F9 pyoverdine, safracin, and phenazine mutants remain effective antagonists against *Erwinia amylovora* in Apple flowers. Appl Environ Microbiol 2020; 86(8): e02620-19.
[http://dx.doi.org/10.1128/AEM.02620-19] [PMID: 32033956]

[168] Bender CL, Alarcón-Chaidez F, Gross DC. *Pseudomonas syringae* phytotoxins: mode of action, regulation, and biosynthesis by peptide and polyketide synthetases. Microbiol Mol Biol Rev 1999; 63(2): 266-92.
[http://dx.doi.org/10.1128/MMBR.63.2.266-292.1999] [PMID: 10357851]

[169] Bensaci MF, Gurnev PA, Bezrukov SM, Takemoto JY. Fungicidal activities and mechanisms of action of *Pseudomonas syringae* pv. *syringae* lipodepsipeptide syringopeptins 22A and 25A. Front Microbiol 2011; 2: 216.
[http://dx.doi.org/10.3389/fmicb.2011.00216] [PMID: 22046175]

[170] Grgurina I, Bensaci M, Pocsfalvi G, *et al.* Novel cyclic lipodepsipeptide from *Pseudomonas syringae* pv. *lachrymans* strain 508 and syringopeptin antimicrobial activities. Antimicrob Agents Chemother 2005; 49(12): 5037-45.
[http://dx.doi.org/10.1128/AAC.49.12.5037-5045.2005] [PMID: 16304170]

[171] Scholz-Schroeder BK, Hutchison ML, Grgurina I, Gross DC. The contribution of syringopeptin and syringomycin to virulence of *Pseudomonas syringae* pv. *syringae* strain B301D on the basis of sypA and syrB1 biosynthesis mutant analysis. Mol Plant Microbe Interact 2001; 14(3): 336-48.
[http://dx.doi.org/10.1094/MPMI.2001.14.3.336] [PMID: 11277431]

[172] Matilla MA, Nogellova V, Morel B, Krell T, Salmond GPC. Biosynthesis of the acetyl-CoA carboxylase-inhibiting antibiotic, andrimid in *Serratia* is regulated by Hfq and the LysR-type transcriptional regulator, AdmX. Environ Microbiol 2016; 18(11): 3635-50.
[http://dx.doi.org/10.1111/1462-2920.13241] [PMID: 26914969]

[173] Liu X, Fortin PD, Walsh CT. Andrimid producers encode an acetyl-CoA carboxyltransferase subunit resistant to the action of the antibiotic. Proc Natl Acad Sci USA 2008; 105(36): 13321-6.
[http://dx.doi.org/10.1073/pnas.0806873105] [PMID: 18768797]

[174] Thistlethwaite IRG, Bull FM, Cui C, *et al.* Elucidation of the relative and absolute stereochemistry of the kalimantacin/batumin antibiotics. Chem Sci (Camb) 2017; 8(9): 6196-201.
[http://dx.doi.org/10.1039/C7SC01670K] [PMID: 28989652]

[175] Aamir M, Rai KK, Zehra A, Dubey MK. Microbial bioformulation-based plant biostimulants: a plausible approach toward next generation of sustainable agriculture. Microbial Endophytes 2020; 195-25.
[http://dx.doi.org/10.1016/B978-0-12-819654-0.00008-9]

[176] Zhang F, Shen J, Jing J, Li L, Chen X. Rhizosphere Processes and Management for Improving Nutrient Use Efficiency and Crop Productivity. In: Xu J, Huang PM, Eds. Molecular Environmental Soil Science at the Interfaces in the Earth's Critical Zone. Berlin, Heidelberg: Springer 2010.
[http://dx.doi.org/10.1007/978-3-642-05297-2_16]

[177] Skraly FA, Ambavaram MMR, Peoples O, Snell KD. Metabolic engineering to increase crop yield: From concept to execution. Plant Sci 2018; 273: 23-32.
[http://dx.doi.org/10.1016/j.plantsci.2018.03.011] [PMID: 29907305]

[178] Bhatt P, Verma A, Verma S, *et al.* Understanding Phytomicrobiome: A potential reservoir for better crop management. Sustainability (Basel) 2020; 12(13): 5446.
[http://dx.doi.org/10.3390/su12135446]

CHAPTER 4

Microbial Communication: A Significant Approach to Understanding Microbial Activities and Interactions

Samia Khanum[1,*], Abdel Rahman M. Tawaha[2], Abdel Razzaq Al-Tawaha[3], Hiba Alatrash[4], Abdur Rauf[5], Arun Karnwal[6], Abhijit Dey[7], Nujoud Alimad[8], Sameena Lone[9], Khursheed Hussain[9], Imran[10], Amanullah[10], Shah Khalid[10], Palani Saranraj[11] and Abdul Basit[12]

[1] *Department of Botany, University of the Punjab, Lahore, Pakistan*

[2] *Department of Biological Sciences, Al Hussein bin Talal University, P.O. Box 20, Maan, Jordan*

[3] *Department of Biological Sciences, Al Hussein bin Talal University, P.O. Box 20, Maan, Jordan*

[4] *General Commission for Scientific Agricultural Research, Syria*

[5] *Department of Chemistry, University of Swabi, Anbar, Khyber Pakhtunkhwa, Pakistan*

[6] *Department of Microbiology, School of Bioengineering and BioSciences, Lovely Professional University, Phagwara, India*

[7] *Department of Life Sciences, Presidency University, Kolkata, India*

[8] *Damascus University, Faculty of Agriculture, Damascus, Syria*

[9] *Division of Vegetable Science, SKUAST-Kashmir, India*

[10] *Department of Agronomy, The University of Agriculture, Peshawar, Pakistan*

[11] *Department of Microbiology, Sacred Heart College (Autonomous), Tirupattur 635601, Tamil Nadu, India*

[12] *Department of Plant Pathology, Agriculture College, Guizhou University, Guiyang 550025, P.R. China*

Abstract: To understand the interaction between different microbes, it is important to understand how they communicate with one another in their adjacent environment. These interactions are beneficial because when different microbes interact, they stimulate specific mechanisms, release signals, and result in the production and synthesis of important vaccines, anti-bacterial and anti-fungal agents, and secondary metabolites. These metabolites are beneficial from a medicinal point of view as well.

* **Corresponding author Samia Khanum:** Department of Botany, University of the Punjab, Lahore, Pakistan; E-mail: samiaghani33@gmail.com

Arun Karnwal & Abdel Rahman Mohammad Said Al-Tawaha (Eds.)

Many studies proved that specific metabolites are released only when they interact with other microorganisms in their adjacent environment. This is also proved through chromatography and co-culturing of these microorganisms.

Keywords: Anti-bacterial, Anti-microbial, Co-culturing of microbes, Microbial metabolites, Medicinal benefits.

INTRODUCTION

The rhizosphere is a complex, nutrient-rich soil with a diversified ecosystem of bacteria, fungus, protists, nematodes, and other microbial communities that dwell around plant roots. A new variety of secondary metabolites produced by bacteria and fungi have fascinated researchers in recent years [1 - 6]. Microorganisms communicate, interact, and live in close associations [7, 8]. The chemical interaction between microorganisms depends on naturally produced signaling molecules which help in communication, interaction, competition and building different defense mechanisms relevant to their habitat [9]. There are two kinds of competition among microbes within an ecological system: interference and scramble between or within species. When one microorganism takes all of the nutrients from another, it is called interference competition, but when nutrients are consumed by one, it is called scramble competition, which is more intense when nutrients are scarce. Different kinds of sustainable secondary metabolites can be produced by the co-culture of microorganisms. Co-culture with transformation, genome mining, and unculturable microorganisms are considered important sources for producing novel antibiotics [10, 11]. Secondary metabolites are the key to controlling other pathogens and inhibiting the growth of competitors [12]. Many studies reveal different co-culturing combinations like fungal-bacterial, fungal-fungal, and bacterial-bacterial associations and their positive impact on producing new compounds. Signaling molecules facilitate the effective co-cultivation experiments in which fungal mycelium interacts with bacterial filaments. Such physical interactions were also observed between *Aspergillus nidulans* and Streptomyces hygroscopicus, and it is observed that the silent gene of *A. nudilans* for secondary metabolites is activated only when interacted with bacterium filaments [13]. It supports the concept that other than diffusible molecules, such physical connection is also part of the induction of cryptic biosynthesis genes. When fungi co-culture with other fungal species, they positively impact the production of metabolites when interacting physically [14, 15]. This also modifies the morphology of the mycelium, secondary metabolites, and extracellular enzymes [16 - 18]. Many sub-elements of the fungal domain can be used to synthesize other products and for many other applications [19]. So by co-culture techniques of various microorganisms, we can obtain many beneficial secondary metabolites.

Bacterial and Fungal Co-Culture

The role of bacterial and fungal co-culturing in the induction of secondary metabolites has been approved by various studies [20]. Marine fungus Pestalotia, co-cultured with marine bacterial strain CNJ-328, produced a chlorinated benzophenone chemical named pestalone, having anti-bacterial qualities against many bacterial infections [21]. When marine bacterium α-proteobacterium (Strain CNJ-328) co-cultured with fungus Libertella sp., it produced diterpenoids, libertellenones A–D (2–5) [22]. Terpenoids are metabolites produced by plants and fungi but very rarely produced by bacteria [23, 24]; that is why it has been proved that these diterpenoids are produced by fungal strains. The biosynthetic cluster gene for the synthesis of compounds 2-5 may be activated in the presence of bacterial strain. *Aspergillus fumigatus,* when co-cultured with many *Streptomyces* spp., helps in the biosynthesis of formyl xanthocillin analogs [25]. When *S. bullii* derived from soil and isolated from the Atacama Desert of South America was co-cultivated with *A. fumigatus,* it produced ten compounds [26].

A new compound, polyketides fumicyclines A and B, which is a fungal derivative yielded from mixed fermentation of *A. fumigatus* and *Streptomyces rapamycinicus* [27], proved that close interaction between fungus and bacterium is required for the synthesis of fungal metabolites. Through studies, it is concluded that the expression of the gene present in *A. fumigatus* changes by *S. rapamycinicus* by modifying regulatory processes. The interactions between *A. nidulans* and *S. hygroscopicus* are responsible for activating fungal gene expression to synthesize polyketide metabolites [13]. It is reported that many compounds are isolated but not synthesized by fungal strains [28]. The silent genes in fungal strains are only activated in the presence of a bacterium strain in a co-culture. Many new compounds have been identified and isolated from co-culturing fungi with bacteria [9]. Cytotoxic and anti-bacterial diketopiperazine disulfide: gilonitrin A has been yielded from co-culture of marine derived A. fumigatus and Sphingomonas sp., and this compound shows strong cytotoxic and anti-bacterial activity [29]. Anti-fungal metabolites were also yielded from the co-culture of fungal species with bacterium species [30]. The growth of fungus was observed to be reduced by some bacteria, but some bacteria improved the bioactivity of the fungal species. Studies also explained that under stress conditions, fungi produced more bioactive compounds [31]. The fungal bioactivity increased by bacteria during mixed fermentation which resulted in the production of secondary metabolites by bacteria acting as signaling molecules which help in the biosynthesis of anti-bacterial compounds [31]. The molecular study of the interaction between *Pseudomonas aeruginosa* and *A. fumigatus* with the help of matrix-assisted laser desorption/ionization-imaging mass spectrometry (MALDI–IMS) combined with tandem mass spectrometry (MS/MS) shows the characteristics of the secondary metabolites excreted by these microorganisms when grown on agar medium. Metabolites produced by one fungal species can be modified by others, resulting in increased toxicity and biosynthesis of fungal siderophore [32, 33]. Mixed fermentation of *Bacillus subtilis* with endophytic fungus Chaetomium sp. increased the production of metabolites up to 8.3-fold [34].

Bacterial-Bacterial Co-Culturing

Studies about interactions with different bacteria have been reported since competition between different Streptomyces species has been detected in the soil [35]. In sterile and improved soil, the survival of *S. dusseldrof* was induced by *S. lividans* but inhibited by *S. bikiniensis* [35]. Mixed fermentation of these bacteria released red-colored pigment showing the synthesis of secondary metabolites in the culture (36). Alchivemycin antibiotic was also biosynthesized in mixed fermentation of *S. endus* and *T. pulmonis* [36]. Bacterium *Streptomyces tenjimariensis* derived from marine synthesized the antibiotics istamycins A and B [37]. The effect of marine bacteria on the production of istamycin by *S. tenjimariensis* has been studied through co-cultivation experimentation [37]. About 12 to 3 bacterial species increase the production of istamycin up to 2 folds in mixed culture than in mono-culture. Istamycin also inhibits the growth of competitive bacterial colonies [37]. Unique secondary metabolites with five actinomycetes have been identified from the co-culture of *S. coelicolor* [38]. In order to detect these metabolites, two techniques were combined namely, nanospray desorption electrospray ionization (Nano-DESI) and matrix-assisted laser desorption ionization-time of flight (MALDI-TOF) imaging mass spectrometry (IMS) [30, 32, 38]. Various bacterial interactions have reported many unidentified secondary metabolites from S. coelicolor co-culture. Secondary metabolites have been produced with high complexity and specificity from the interaction between actinomycetes bacteria (38). The interaction between *S. coelicolor* and five actinomycetes induced acyl-desferrioxamine siderophores' production [38]. Various studies and experimentations have proved that co-culturing of bacterial species results in different kinds of secondary metabolites, antibiotics, cytotoxic alkaloids, and many other essential molecules not produced in mono-cultures [39 - 41].

Fungal-Fungal Co-Culturing

Mixed fermentation of fungal species derived from sediments yielded compounds with cytotoxic activity against many tumor cells [42]. These compounds also show anti-bacterial activity against many bacteria [42]. The plant endophytic fungi, when co-cultured, produced linear depsipeptides subenniatins A and B and also produced cyclic depsipeptides enniantis A, A1, B, and B1 [43]. Many volatile compounds have been delivered when *Candida albicans* are co-cultured with skin-commensal fungi [44]. It is reported that dihydrofarnesol is only produced in the presence of *T. rubrum* and has more anti-fungal activity than other anti-fungal agents [44]. Two mangroves-derived epiphytic *Aspergillus* species produced new alkaloids, neoaspergillic acid, and ergosterol when co-cultured [45]. Co-culturing *Trametes versicolor, Bjerkandera adusta,* and *Hypholoma fasciculare* produced various essential metabolites. The interaction between mycelia has influenced the expression of biosynthetic pathways [46]. *T. versicolor* produced pharmaceutically significant metabolites [46, 47]. Soil-derived fungi, when co-cultured produced secopenicillide C, penicillide, MC-141, stromemycin, and pestalasin A [48]. It has been proved through various studies that such important metabolites are not produced in mono-cultures.

Biological Importance of Co-Culture

It is proved that multiple microorganisms, when co-cultivated, result in improved biosynthesis of important bioactive metabolites [49]. New drugs can also be synthesized by mixed fermentation [50]. The most common example of this is an antagonistic response of marine bacteria co-cultivation and successful interaction with terrestrial bacteria. Studies show that marine epibiotic bacteria produced many antibiotic agents because of antagonistic interaction with pathogenic bacteria [51].

Analytical Methods for Identification of Metabolites from Co-Culture

In mixed fermentation experimentation, the most critical point is identifying and comparing metabolites from mono and co-cultures. High-Performance Liquid Chromatography (HPLC) is one of the most important techniques to monitor the change between chromatograms of mono and co-cultures and detect any change in metabolic profiles [33]. This is the way to identify metabolites of co-culture quickly. Chromatographic techniques, including semi-preparative HPLC, Counter-Current Chromatography (CCC), Over-Pressured Layer Chromatography (OPLC), and Preparative Thin-layer Chromatography (PTLC), are the techniques used for the purification of metabolites for structural analysis [52, 53]. Software-oriented semi-preparative HPLC-MS was used to purify compounds from fungal co-culture previously. Nano-Electrospray Ionization (nanoESI) [54] and Chip-Based nanoESI techniques are the best techniques for the identification of metabolites [55 - 59]. Ultra-High Pressure Liquid Chromatography Time of Flight (UHPLC-TOF) coupled to a micro-TOF-2Q mass spectrometer (MS) with an ESI interface is the most efficient technique used recently for the identification of metabolites in both positive and negative ion modes in co-culture. Great achievements have been made by using the MALDI-IMS technique recently because it is beneficial for the identification of secondary metabolites [30, 32, 33]. MALDI-TOF and MALDI-FT-ICR imaging mass spectrometry (MALDI-IMS), combined with MS/MS networking, have been used to study the interaction between species and identify compounds from microbial extracts [30, 32]. Nuclear Magnetic Resonance (NMR) spectroscopy is used for the characterization of metabolites, and this technique is based on metabolism combined with multivariate data analysis [60]. This technique helps in direct biochemical analysis of the metabolites [61]. NMR spectrometers with high resolution were used for 1D and 2D experimentation [62]. NMR probes, when cryogenically cooled, can measure small samples as tens of micrograms of purified substance [62].

CONCLUSION

It is vital to consider how microbes interact with one another in their respective environments. These interactions are constructive, as they activate specific pathways, release signals and lead to essential vaccines, anti-bacterial and anti-fungal agents, and to the production and synthesis of secondary metabolites. These metabolites are also beneficial for medicinal purposes. Several studies have shown that such metabolites are only released when interacting with other microorganisms in their neighboring settings.

CONSENT FOR PUBLICATION

Not applicable.

CONFLICT OF INTEREST

The author declares no conflict of interest, financial or otherwise.

ACKNOWLEDGEMENT

Declared none.

REFERENCES

[1]   Abdalla MA, Helmke E, Laatsch H. Bioactive Angucyclinone from a Marine Derived Streptomyces sp. B6219. 2010.

[2]   Jumpathong J, Abdalla MA, Lumyong S, Laatsch H. Stemphol galactoside, a new stemphol derivative isolated from the tropical endophytic fungus Gaeumannomyces amomi. Natural Product Communications 2010; 5(4): 1934578X1000500415.

[3]   Abdalla MA, Yadav PP, Dittrich B, Schüffler A, Laatsch H. ent-Homoabyssomicins A and B, two new spirotetronate metabolites from Streptomyces sp. Ank 210. Org Lett 2011; 13(9): 2156-9.
[http://dx.doi.org/10.1021/ol103076y] [PMID: 21446696]

[4]   Abdalla MA, Win HY, Islam MT, von Tiedemann A, Schüffler A, Laatsch H. Khatmiamycin, a
        motility inhibitor and zoosporicide against the grapevine downy mildew pathogen Plasmopara viticola
        from Streptomyces sp. ANK313. J Antibiot (Tokyo) 2011; 64(10): 655-9.
        [http://dx.doi.org/10.1038/ja.2011.68] [PMID: 21811263]

[5]   Zinad DS, Shaaban KA, Abdalla MA, Islam MT, Schüffler A, Laatsch H. Bioactive isocoumarins
        from a terrestrial Streptomyces sp. ANK302. Natural Product Communications 2011; 6(1):
        1934578X1100600111.
        [http://dx.doi.org/10.1177/1934578X1100600111]

[6]   Abdalla MA, Matasyoh JC. Endophytes as producers of peptides: an overview about the recently
        discovered peptides from endophytic microbes. Nat Prod Bioprospect 2014; 4(5): 257-70.
        [http://dx.doi.org/10.1007/s13659-014-0038-y] [PMID: 25205333]

[7]   Strobel G, Daisy B. Bioprospecting for microbial endophytes and their natural products. Microbiol
        Mol Biol Rev 2003; 67(4): 491-502.
        [http://dx.doi.org/10.1128/MMBR.67.4.491-502.2003] [PMID: 14665674]

[8]   Aly AH, Debbab A, Proksch P. Fungal endophytes: unique plant inhabitants with great promises. Appl
        Microbiol Biotechnol 2011; 90(6): 1829-45.
        [http://dx.doi.org/10.1007/s00253-011-3270-y] [PMID: 21523479]

[9]   Ola ARB, Thomy D, Lai D, Brötz-Oesterhelt H, Proksch P. Inducing secondary metabolite production
        by the endophytic fungus Fusarium tricinctum through coculture with *Bacillus subtilis*. J Nat Prod
        2013; 76(11): 2094-9.
        [http://dx.doi.org/10.1021/np400589h] [PMID: 24175613]

[10]   Wilson MC, Mori T, Rückert C, *et al.* An environmental bacterial taxon with a large and distinct
        metabolic repertoire. Nature 2014; 506(7486): 58-62.
        [http://dx.doi.org/10.1038/nature12959] [PMID: 24476823]

[11]   Ling LL, Schneider T, Peoples AJ, *et al.* A new antibiotic kills pathogens without detectable
        resistance. Nature 2015; 517(7535): 455-9.
        [http://dx.doi.org/10.1038/nature14098] [PMID: 25561178]

[12]   Davies J. What are antibiotics? Archaic functions for modern activities. Mol Microbiol 1990; 4(8):
        1227-32.
        [http://dx.doi.org/10.1111/j.1365-2958.1990.tb00701.x] [PMID: 2280684]

[13]   Schroeckh V, Scherlach K, Nützmann HW, *et al.* Intimate bacterial–fungal interaction triggers
        biosynthesis of archetypal polyketides in *Aspergillus nidulans*. Proc Natl Acad Sci USA 2009;
        106(34): 14558-63.
        [http://dx.doi.org/10.1073/pnas.0901870106] [PMID: 19666480]

[14]   Li C, Zhang J, Shao C, Ding W, She Z, Lin Y. A new xanthone derivative from the co-culture broth of
        two marine fungi (strain No. E33 and K38). Chem Nat Compd 2011; 47(3): 382-4.
        [http://dx.doi.org/10.1007/s10600-011-9939-8]

[15]   Zhu F, Chen G, Wu J, Pan J. Structure revision and cytotoxic activity of marinamide and its methyl
        ester, novel alkaloids produced by co-cultures of two marine-derived mangrove endophytic fungi. Nat
        Prod Res 2013; 27(21): 1960-4.
        [http://dx.doi.org/10.1080/14786419.2013.800980] [PMID: 23701463]

[16]   Griffith GS, Rayner ADM, Wildman HG. Interspecific interactions, mycelial morphogenesis and
        extracellular metabolite production in Phlebia radiata (Aphyllophorales). Nova Hedwigia 1994; 59(3):
        331-44.

[17]   Rayner AD, Griffith GS, Wildman HG. Induction of metabolic and morphogenetic changes during
        mycelial interactions among species of higher fungi, 1994.
        [http://dx.doi.org/10.1042/bst0220389]

[18]   Woodward S, Boddy L. Interactions between saprotrophic fungi. Ecology of saprotrophic

Basidiomycetes 2008; 125-41.

[19]   Frey-Klett P, Burlinson P, Deveau A, Barret M, Tarkka M, Sarniguet A. Bacterial-fungal interactions: hyphens between agricultural, clinical, environmental, and food microbiologists. Microbiol Mol Biol Rev 2011; 75(4): 583-609.
[http://dx.doi.org/10.1128/MMBR.00020-11] [PMID: 22126995]

[20]   Abdalla MA, Sulieman S, McGaw LJ. Microbial communication: A significant approach for new leads. S Afr J Bot 2017; 113: 461-70.
[http://dx.doi.org/10.1016/j.sajb.2017.10.001]

[21]   Cueto M, Jensen PR, Kauffman C, Fenical W, Lobkovsky E, Clardy J. Pestalone, a new antibiotic produced by a marine fungus in response to bacterial challenge. J Nat Prod 2001; 64(11): 1444-6.
[http://dx.doi.org/10.1021/np0102713] [PMID: 11720529]

[22]   Oh DC, Jensen PR, Kauffman CA, Fenical W. Libertellenones A–D: Induction of cytotoxic diterpenoid biosynthesis by marine microbial competition. Bioorg Med Chem 2005; 13(17): 5267-73.
[http://dx.doi.org/10.1016/j.bmc.2005.05.068] [PMID: 15993608]

[23]   Turner WB, Aldridge DC. Fungal Metabolites II. London: Academic Press 1983; p. 631.

[24]   Hefter J, Richnow HH, Fischer U, Trendel JM, Michaelis W. (-)-Verrucosan-2β-ol from the phototrophic bacterium Chloroflexus aurantiacus: first report of a verrucosane-type diterpenoid from a prokaryote. Microbiology 1993; 139(11): 2757-61.

[25]   Zuck KM, Shipley S, Newman DJ. Induced production of N-formyl alkaloids from Aspergillus fumigatus by co-culture with Streptomyces peucetius. J Nat Prod 2011; 74(7): 1653-7.
[http://dx.doi.org/10.1021/np200255f] [PMID: 21667925]

[26]   Rateb ME, Hallyburton I, Houssen WE, et al. Induction of diverse secondary metabolites in Aspergillus fumigatus by microbial co-culture. RSC Advances 2013; 3(34): 14444-50.
[http://dx.doi.org/10.1039/c3ra42378f]

[27]   König CC, Scherlach K, Schroeckh V, et al. Bacterium induces cryptic meroterpenoid pathway in the pathogenic fungus Aspergillus fumigatus. ChemBioChem 2013; 14(8): 938-42.
[http://dx.doi.org/10.1002/cbic.201300070] [PMID: 23649940]

[28]   Sato S, Morishita T, Hosoya T, Ishikawa Y. Novel pentacyclic compounds, F-9775A and F-9775B, their manufacture with Paecilomyces carneus, and their use for treatment of osteoporosis. Japanese patent JP11001480, 1999.

[29]   Park HB, Kwon HC, Lee CH, Yang HO. Glionitrin A, an antibiotic-antitumor metabolite derived from competitive interaction between abandoned mine microbes. J Nat Prod 2009; 72(2): 248-52.
[http://dx.doi.org/10.1021/np800606e] [PMID: 19159274]

[30]   Moree WJ, Yang JY, Zhao X, et al. Imaging mass spectrometry of a coral microbe interaction with fungi. J Chem Ecol 2013; 39(7): 1045-54.
[http://dx.doi.org/10.1007/s10886-013-0320-1] [PMID: 23881443]

[31]   Miao L, Kwong TFN, Qian PY. Effect of culture conditions on mycelial growth, antibacterial activity, and metabolite profiles of the marine-derived fungus Arthrinium c.f. saccharicola. Appl Microbiol Biotechnol 2006; 72(5): 1063-73.
[http://dx.doi.org/10.1007/s00253-006-0376-8] [PMID: 16538484]

[32]   Moree WJ, Phelan VV, Wu CH, et al. Interkingdom metabolic transformations captured by microbial imaging mass spectrometry. Proc Natl Acad Sci USA 2012; 109(34): 13811-6.
[http://dx.doi.org/10.1073/pnas.1206855109] [PMID: 22869730]

[33]   Rutledge PJ, Challis GL. Discovery of microbial natural products by activation of silent biosynthetic gene clusters. Nat Rev Microbiol 2015; 13(8): 509-23.
[http://dx.doi.org/10.1038/nrmicro3496] [PMID: 26119570]

[34]   Akone SH, Mándi A, Kurtán T, et al. Inducing secondary metabolite production by the endophytic

fungus Chaetomium sp. through fungal–bacterial co-culture and epigenetic modification. Tetrahedron 2016; 72(41): 6340-7.
[http://dx.doi.org/10.1016/j.tet.2016.08.022]

[35]   Turpin PE, Dhir VK, Maycroft KA, Rowlands C, Wellington EMH. The effect of Streptomyces species on the survival of Salmonella in soil. FEMS Microbiol Ecol 1992; 10(4): 271-80.
[http://dx.doi.org/10.1111/j.1574-6941.1992.tb01664.x]

[36]   Onaka H, Mori Y, Igarashi Y, Furumai T. Mycolic acid-containing bacteria induce natural-product biosynthesis in Streptomyces species. Appl Environ Microbiol 2011; 77(2): 400-6.
[http://dx.doi.org/10.1128/AEM.01337-10] [PMID: 21097597]

[37]   Slattery M, Rajbhandari I, Wesson K. Competition-mediated antibiotic induction in the marine bacterium Streptomyces tenjimariensis. Microb Ecol 2001; 41(2): 90-6.
[http://dx.doi.org/10.1007/s002480000084] [PMID: 12032613]

[38]   Traxler MF, Watrous JD, Alexandrov T, Dorrestein PC, Kolter R. Interspecies interactions stimulate diversification of the Streptomyces coelicolor secreted metabolome. MBio 2013; 4(4): e00459-13.
[http://dx.doi.org/10.1128/mBio.00459-13] [PMID: 23963177]

[39]   Dashti Y, Grkovic T, Abdelmohsen U, Hentschel U, Quinn R. Production of induced secondary metabolites by a co-culture of sponge-associated actinomycetes, Actinokineospora sp. EG49 and Nocardiopsis sp. RV163. Mar Drugs 2014; 12(5): 3046-59.
[http://dx.doi.org/10.3390/md12053046] [PMID: 24857962]

[40]   Wiener P. Experimental studies on the ecological role of antibiotic production in bacteria. Evol Ecol 1996; 10(4): 405-21.
[http://dx.doi.org/10.1007/BF01237726]

[41]   Hoshino S, Zhang L, Awakawa T, Wakimoto T, Onaka H, Abe I. Arcyriaflavin E, a new cytotoxic indolocarbazole alkaloid isolated by combined-culture of mycolic acid-containing bacteria and Streptomyces cinnamoneus NBRC 13823. J Antibiot (Tokyo) 2015; 68(5): 342-4.
[http://dx.doi.org/10.1038/ja.2014.147] [PMID: 25335694]

[42]   Pettit R, Pettit G, Xu JP, Weber C, Richert L. Isolation of human cancer cell growth inhibitory, antimicrobial lateritin from a mixed fungal culture. Planta Med 2010; 76(5): 500-1.
[http://dx.doi.org/10.1055/s-0029-1240617] [PMID: 19941263]

[43]   Wang J, Lin W, Wray V, Lai D, Proksch P. Induced production of depsipeptides by co-culturing Fusarium tricinctum and Fusarium begoniae. Tetrahedron Lett 2013; 54(20): 2492-6.
[http://dx.doi.org/10.1016/j.tetlet.2013.03.005]

[44]   Brasch J, Horter F, Fritsch D, Beck-Jendroschek V, Tröger A, Francke W. Acyclic sesquiterpenes released by Candida albicans inhibit growth of dermatophytes. Med Mycol 2014; 52(1): 46-55.
[PMID: 23902158]

[45]   Zhu F, Chen G, Chen X, Huang M, Wan X. Aspergicin, a new antibacterial alkaloid produced by mixed fermentation of two marine-derived mangrove epiphytic fungi. Chem Nat Compd 2011; 47(5): 767-9.
[http://dx.doi.org/10.1007/s10600-011-0053-8]

[46]   Eyre C, Muftah W, Hiscox J, et al. Microarray analysis of differential gene expression elicited in Trametes versicolor during interspecific mycelial interactions. Fungal Biol 2010; 114(8): 646-60.
[http://dx.doi.org/10.1016/j.funbio.2010.05.006] [PMID: 20943176]

[47]   Zjawiony JK. Biologically active compounds from Aphyllophorales (polypore) fungi. J Nat Prod 2004; 67(2): 300-10.
[http://dx.doi.org/10.1021/np030372w] [PMID: 14987072]

[48]   Nonaka K, Abe T, Iwatsuki M, et al. Enhancement of metabolites productivity of Penicillium pinophilum FKI-5653, by co-culture with Trichoderma harzianum FKI-5655. J Antibiot (Tokyo) 2011; 64(12): 769-74.

[http://dx.doi.org/10.1038/ja.2011.91] [PMID: 22008698]

[49]   Ueda K, Beppu T. Antibiotics in microbial coculture. J Antibiot (Tokyo) 2017; 70(4): 361-5.
[http://dx.doi.org/10.1038/ja.2016.127] [PMID: 27756913]

[50]   Marmann A, Aly A, Lin W, Wang B, Proksch P. Co-cultivation-a powerful emerging tool for enhancing the chemical diversity of microorganisms. Mar Drugs 2014; 12(2): 1043-65.
[http://dx.doi.org/10.3390/md12021043] [PMID: 24549204]

[51]   Burgess JG, Jordan EM, Bregu M, Mearns-Spragg A, Boyd KG. Microbial antagonism: a neglected avenue of natural products research. J Biotechnol 1999; 70(1-3): 27-32.
[http://dx.doi.org/10.1016/S0168-1656(99)00054-1] [PMID: 10412203]

[52]   Klebovich I, Mincsovics E, Szúnyog J, *et al.* Isolation and identification of metabolites of 3H-and 14C-deramciclane by OPLC-digital autoradiography on-line sample collection and mass spectrometry. J Planar Chromatogr Mod TLC 1998; 11(5): 394-9.

[53]   Do TKT, Hadji-Minaglou F, Antoniotti S, Fernandez X. Secondary metabolites isolation in natural products chemistry: Comparison of two semipreparative chromatographic techniques (high pressure liquid chromatography and high performance thin-layer chromatography). J Chromatogr A 2014; 1325: 256-60.
[http://dx.doi.org/10.1016/j.chroma.2013.11.046] [PMID: 24377738]

[54]   Schmidt A, Karas M, Dülcks T. Effect of different solution flow rates on analyte ion signals in nano-ESI MS, or: when does ESI turn into nano-ESI? J Am Soc Mass Spectrom 2003; 14(5): 492-500.
[http://dx.doi.org/10.1016/S1044-0305(03)00128-4] [PMID: 12745218]

[55]   Lazar IM, Grym J, Foret F. Microfabricated devices: A new sample introduction approach to mass spectrometry. Mass Spectrom Rev 2006; 25(4): 573-94.
[http://dx.doi.org/10.1002/mas.20081] [PMID: 16508917]

[56]   Wickremsinhe E, Singh G, Ackermann B, Gillespie T, Chaudhary A. A review of nanoelectrospray ionization applications for drug metabolism and pharmacokinetics. Curr Drug Metab 2006; 7(8): 913-28.
[http://dx.doi.org/10.2174/138920006779010610] [PMID: 17168691]

[57]   Pereira-Medrano AG, Sterling A, Snijders APL, Reardon KF, Wright PC. A systematic evaluation of chip-based nanoelectrospray parameters for rapid identification of proteins from a complex mixture. J Am Soc Mass Spectrom 2007; 18(9): 1714-25.
[http://dx.doi.org/10.1016/j.jasms.2007.06.011] [PMID: 17689093]

[58]   Almeida R, Mosoarca C, Chirita M, *et al.* Coupling of fully automated chip-based electrospray ionization to high-capacity ion trap mass spectrometer for ganglioside analysis. Anal Biochem 2008; 378(1): 43-52.
[http://dx.doi.org/10.1016/j.ab.2008.03.039] [PMID: 18406832]

[59]   Lydic TA, Busik JV, Esselman WJ, Reid GE. Complementary precursor ion and neutral loss scan mode tandem mass spectrometry for the analysis of glycerophosphatidylethanolamine lipids from whole rat retina. Anal Bioanal Chem 2009; 394(1): 267-75.
[http://dx.doi.org/10.1007/s00216-009-2717-9] [PMID: 19277613]

[60]   Wu C, Kim HK, van Wezel GP, Choi YH. Metabolomics in the natural products field – a gateway to novel antibiotics. Drug Discov Today Technol 2015; 13: 11-7.
[http://dx.doi.org/10.1016/j.ddtec.2015.01.004] [PMID: 26190678]

[61]   Kim HK, Choi YH, Verpoorte R. NMR-based metabolomic analysis of plants. Nat Protoc 2010; 5(3): 536-49.
[http://dx.doi.org/10.1038/nprot.2009.237] [PMID: 20203669]

[62]   Simpson JH. Organic structure determination using 2-D NMR spectroscopy: a problem-based approach. Academic Press 2011.

CHAPTER 5

Nutrient Cycling: An Approach for Environmental Sustainability

Sufiara Yousuf[1], Nafiaah Naqash[1] and **Rahul Singh[1,*]**

[1] *Department of Zoology, School of Bio-engineering and Bio-sciences, Lovely Professional University, Phagwara-144411, Punjab, India*

Abstract: Nutrient cycling is an important environmental process and has been the focus of ecological research. Nutrient cycling refers to the sufficient supply of key elements provided through the ecological processes within and between various biotic or abiotic components of a cell, community, or ecosystem. Nutrient cycling also includes the recovery and reuse of industrial, agricultural, and municipal organic debris that are considered wastes. Nutrient cycles include biotic and abiotic components involved in biological, geological, and chemical processes known as biogeochemical cycles. Changes occurring in such cycles may indicate or even alter the functioning of the ecosystem. Plants take up soil nutrients in terrestrial ecosystems for healthy growth and development, wherein soil acts as a nutrient reservoir. Nutrients are lost from such sites due to soil erosion, denitrification, and food production, which cause reduced availability of nutrients. Therefore, analyzing nutrients' assimilation, transport through biota, and their release for subsequent re-assimilation is mandatory. Nutrients to be recycled essentially for the survival of organisms include macronutrients (C, O, H, N, K, P, Ca, Mg, S, and Cl) and micronutrients (Fe, Mn, Mo, Cu, Zn, Bo, Ni, Co, Na, Se, and I). This chapter presents the role of nutrients and nutrient cycling for environmental sustainability.

Keywords: Anthropogenic Activities, Automobile Emissions, Ammonification, Available Nutrients, Biogeochemical Cycles, Deforestation, Denitrification, Eutrophication, Fossil Fuel, Global Warming, Industrial Wastes, Micro-Nutrients, Macronutrients, Nitrification, Nutrient Cycling, Nutrient inputs, Organic Matter, Soil erosion, Synthetic Fertilizers, Vermicompost.

INTRODUCTION

Nutrients are essential for the growth of living organisms. Nutrients exist in gaseous form (such as O_2, N_2, and CO_2), mineral form (such as apatite, mainly P-containing mineral), inorganic ionic form (such as NH_4^+, NO_3^-, SO_4^{2-}), and organic

---

[*] **Corresponding author Rahul Singh:** Department of Zoology, School of Bioengineering and Biosciences, Lovely Professional University, Phagwara-144411, Punjab, India; E-mail: rahulsingh.mlkzoology@gmail.com

form (mainly C- containing compounds in living or dead organisms or their products). Plants obtain inorganic nutrients from the environment in the form of simple compounds. For example, plants acquire carbon in gaseous form as carbon dioxide, nitrogen in the form of ions as ammonium or nitrate, and Phosphorus as phosphate ions. Plants absorb the dissolved form of these ions from soil water and utilize them in photosynthesis and other metabolic processes. In comparison, heterotrophic organisms acquire these nutrients by consuming food from plant biomass or other heterotrophs. The ingested food contains organic forms of nutrients which can be digested by the animals in their gut, assimilating the inorganic or simple organic forms. The heterotrophs utilize these compounds to carry out various metabolic processes, while micro-organisms take nutrients in mineral form or organic form by decomposition of plant and animal remains (Fig. 1). This figure defines the four major compartments, including atmosphere, rocks and soils, available nutrients, and organic matter. The atmosphere comprises gases and low concentrations of water vapors and particulates. Rocks and soils possess insoluble minerals that are not available directly for uptake by different organisms. The insoluble form of nutrients becomes available for uptake through various chemical transformations such as weathering. Available nutrients are in the form of chemicals and are water-soluble. These nutrients are directly absorbed by the organisms from their surroundings, contributing to their nutrition. Organic matter comprises the nutrients present in living biomass and dead matter. The living biomass can be categorized as autotrophs (autotrophic plants, bacteria, and algae) and heterotrophs (herbivores, carnivores, omnivores, and detritivores).

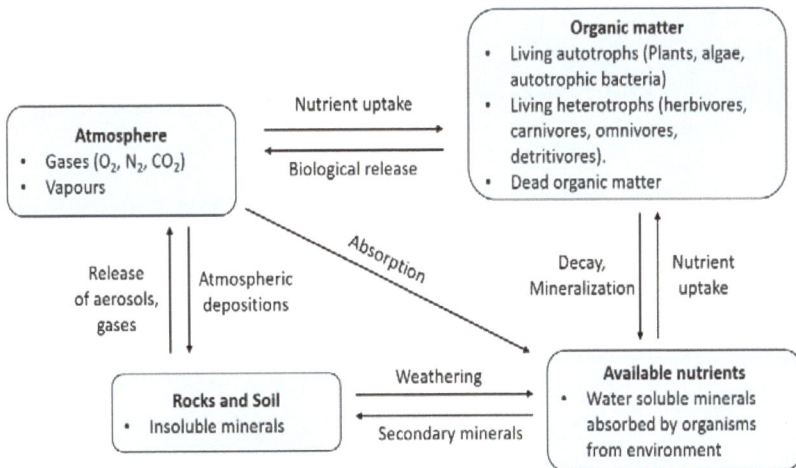

Fig. (1). Cycling of nutrients in different forms between the atmosphere, rocks, soil, and living and dead organic matter.

Nutrients that are required in large quantities are called macronutrients, such as C, O, H, N, K, P, Ca, Mg, S, and Cl. These nutrients nourish the plants and maintain their general state. These nutrients are used by plants in photosynthesis, cellular respiration, enzyme activity, control of stomatal aperture, and electron carriers in chloroplasts and mitochondria. Nutrients, such as Fe, Mn, Mo, Cu, Zn, Bo, Ni, Co, Na, Se, and I, are known as micronutrients. These micronutrients are required to synthesize chlorophyll, and for nitrogen metabolism, nitrogen fixation, protein synthesis, hormone synthesis, tissue support, and sugar transport [1]. When nutrients are stored in the soil for longer, they assure a positive effect on the root's development and growth of the plant, leading to higher crop yields [2]. Nutrients reach the soil from weathering of rocks, atmosphere, and bacterial conversions. In the terrestrial ecosystem, nutrients are available in forests, grasslands, and crop production systems. Nutrient-rich systems exhibit higher biomass production. Nutrients are released from the soils due to soil erosion, conversion of forests into farmland and are discharged into the aquatic environment, which is a common problem, however, the application of Vermicomposting proved an effective solution to this problem. The application of Vermicompost produced from food wastes, paper wastes, and cow dung to soils, results in the addition of soil nutrient contents and microbial biomass [3].

Nutrients moving in a cycle within and between ecosystems and coming back to the environment is known as nutrient cycling. The components of the terrestrial ecosystem above or below the ground depend on each other to carry out these processes [4]. Although the biodiversity of different ecosystems is related to nutrient cycling, the complex process is unclear so far [5, 6]. Nutrients that must be recycled and are essential for the survival of organisms include carbon, hydrogen, oxygen, phosphorus, and nitrogen. These elements are involved in biological, geological, and chemical processes known as biogeochemical cycles. Changes in the biogeochemical cycles can influence the climate.

Nutrient Cycling in Terrestrial Ecosystem

Nutrients are components essential for the growth and development of living organisms. Humans, plants, and animals need nutrients to carry out various metabolic processes such as breathing, repair, *etc*. The introduction of nutrients into the environment makes the ecosystem stable and fertile. Nutrients flow from organism to organism in a circular motion. Nutrient cycling is a natural process and is one of the main ecosystem services to sustain life on earth [7]. Within the terrestrial ecosystem, nutrients are available in the atmosphere, soil, and rocks [8]. Soil acts as a pool of available nutrients that are uptaken by plants for growth and development, released back by direct leaching or from organic matter by biological decomposition. Through this system of intake and output of nutrients, the terrestrial ecosystem participates in different biogeochemical cycles- The carbon cycle, Phosphorus cycle, Nitrogen cycle, and Oxygen cycle.

Importance of Soil

The soil in itself is a dynamic ecosystem rather than being the component of all the terrestrial ecosystems. Soil is developed from the parent material like rocks and minerals, which influences the type of soil developed. Soil formation is also affected by certain climatic factors and biological processes such as temperature and precipitation. The precipitation dissolves minerals which in turn modify the chemistry of soil from the surface to deeper parts. Soil provides a habitat for the immense diversity of living organisms, including micro-organisms [9]. Soil is essential for nutrient cycling as terrestrial plants obtain much of their nutrients and water through their roots. Micro-organisms in the soil also play a crucial role in litter decomposition and nutrient cycling. The micro-organisms oxidize organic matter into water, CO_2, and inorganic nutrients (ammonium).

Soil is also important from an agricultural perspective; it critically influences the growth of crops. Soils are classified into various types depending on the ecological conditions they develop. In cultivable lands, a uniform plow layer is formed at the surface as it is repeatedly mixed up [10]. Due to the passage of repeated machinery, nutrient concentrations, and other practices mandatory for crop productivity, agricultural lands are often deficient in organic matter. The most productive agricultural soils are found along the rivers known as alluvial deposits. These soils are productive as periodic flooding and depositions initiate the abundant supply of nutrients [11].

An insufficient quantity of nutrients may result in less ecological productivity than it could be if present in required quantities. However, excess nutrients may result in environmental issues caused due to toxicity [12]. Cycling nutrients are in the form of organic and inorganic matter within the ecosystem. The processes within nutrient cycles are explained by carbon, nitrogen, phosphorus, and sulfur cycles.

Carbon Cycle

The carbon cycle illustrates the process in which carbon moves from the atmosphere to the earth and back again into the atmosphere. Terrestrial ecosystem acts as a global sink for carbon. Atmospheric CO_2 is taken up by terrestrial ecosystems for photosynthesis and other oxidative processes and is released back when the organisms die, through fire blaze and other mechanisms. Carbon is transmitted from the atmosphere to the soil by 'carbon-fixing' autotrophic organisms, mostly photosynthesizing plants and photo- and chemoautotrophic micro-organisms, which convert atmospheric carbon dioxide (CO_2) into organic material. Fixed carbon is returned to the atmosphere through various mechanisms that account for autotrophic and heterotrophic organisms' respiration. The reverse route involves the decomposition of organic matter by 'organic carbon-consuming' heterotrophic micro-organisms, which use carbon from either plant, animal, or microbial origin as a surface for metabolism, maintaining some carbon in their biomass while releasing the rest as metabolites or CO_2 (Fig. **2**). Bacteria responsible for carbon dioxide release in the atmosphere include *Bacteroides succinogenes*, *Clostridium butyricum,* and *Syntrophomonas* sp.

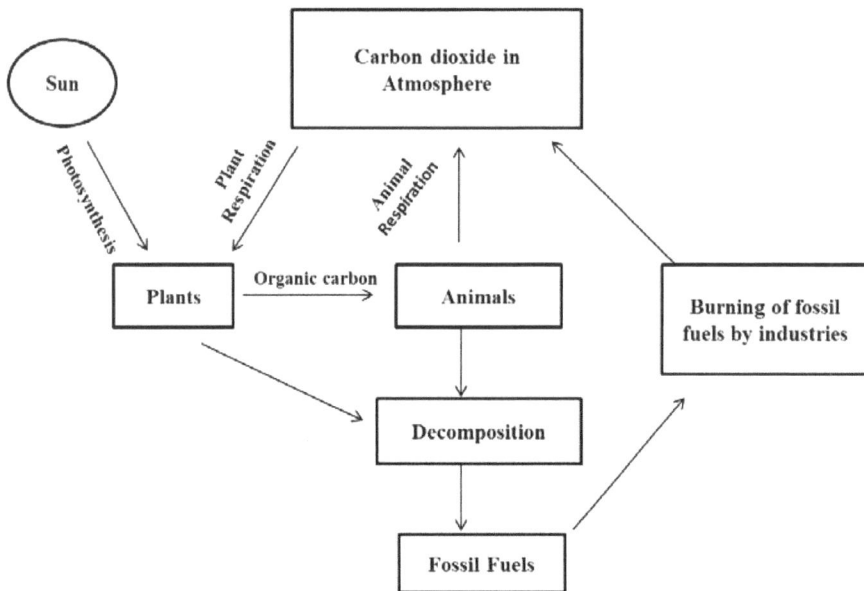

Fig. (2). Carbon cycle.

Atmospheric CO_2 is increasing mainly due to anthropogenic activities like fossil fuel burning, deforestation, and land conversion, which result in climate change and global warming. The emissions of CO_2 and CH_4 are larger as compared to the uptake of these gases. This eventually has resulted in an imbalance, intensifying the greenhouse effect and causing global warming. The Kyoto Protocol (IPCC 2000) was developed to minimize greenhouse gas emissions that cause global warming. It applies to six greenhouse gases: carbon dioxide, methane, nitrogen oxide, perfluorocarbons, hydrofluorocarbons, and sulfur hexafluoride.

Phosphorus Cycle

Phosphorus is essential for life. It is a key component of DNA and RNA and, *via* the high-energy compound, ATP promotes cell metabolism and is an essential component of lipid membranes [13]. In general, the amount of phosphorus in the soil is limited, which also restricts plants' growth. That is why people on farmland also use phosphate fertilizers. By eating plants or plant-eating animals, animals consume phosphates [14]. Phosphorus moves through rocks, soil, organisms, water, and sediments in a cycle.

The phosphorus cycle explains the movement of phosphorus through the ecosystem. The phosphorus cycle differs from other biogeochemical cycles because phosphorus does not exist in the gaseous phase, although phosphoric acid may be found in the atmosphere due to acid rain. The largest pool of phosphorus is found in sedimentary rocks; however, soil micro-organisms also act as a source and sink of phosphorus in the cycle [15]. Due to weathering of rocks, phosphates are circulated throughout the soils and water, which are then taken up by plants for cell development and other processes, released back into the soil by plant-eating animals or by decomposition of plants and animals after death. It has been estimated that 50 million tonnes of phosphorus fertilizers are manufactured per annum. Therefore, it is considered a significant input to the phosphorus cycle as vegetation soil naturally absorbs about 200 million tonnes of phosphorus per annum. The phosphate ion is an effective form of plant phosphorus that is generally present in minor quantities, continuously processed from calcium, phosphate, iron and magnesium. Phosphates are also produced by the process of microbial oxidation of organic phosphorus. Water-soluble phosphorus is absorbed by plant roots and microbes, which are further used in synthesizing a wide range of biochemicals.

Nitrogen Cycle

Nitrogen is considered an important nutrient for living organisms. It is an essential component of proteins, nucleic and amino acids, and chlorophyll. Nitrogen cycle describes how nitrogen is converted into different forms, passing from atmosphere to soil to organisms and recurring back to the atmosphere. Nitrogen occurs in three compartments: the atmosphere, terrestrial organic matter, and marine. The largest source of nitrogen is the earth's atmosphere- 78%, which virtually occurs in the form of nitrogen gas [16]. Other gaseous forms are nitrogen dioxide, nitric oxide, ammonia, and nitrous oxide. The nitrogen cycle includes processes such as Nitrogen fixation, nitrification, ammonification, and denitrification. The process involves the cycling of numerous organic and inorganic nitrogen forms in the ecosystem (Fig. **3**).

Nitrogen Fixation

Atmospheric nitrogen gas is converted into nitrites and nitrates through atmospheric, industrial, and biological processes. The atmospheric process involves the breakdown of nitrogen into nitrogen oxides by lightening, which plants then use. The industrial process involves manufacturing ammonia, which is later converted into fertilizer like Urea. Industrially, about 30% of total fixed nitrogen is produced using Haber- Bosch [17]. Bacteria that carry out the nitrogen fixation process involve *Azotobacter* and Archaea. The species involved in nitrogen fixation are capable of metabolizing nitrogen to ammonia gas. Thereby, ammonia becomes available indirectly to autotrophic plants and micro-organisms incapable of fixing nitrogen themselves. The biological process involves the transformation of nitrogen into ammonia by the bacteria *Rhizobium* and *Azotobacter*. These bacteria live in mutual relationships, in the root nodules of legumes and increase the nitrogen content in nitrogen-poor soils. Certain nitrogen-fixing micro-organisms are free-living, such as cyanobacteria. Compared with biological nitrogen fixation, the minor sources of fixed nitrogen are atmospheric depositions of ammonium and nitrate and uptake of nitric oxide and nitrogen dioxide by plants.

Nitrogen fixation can also occur non-biologically; during lightning, atmospheric nitrogen combines with oxygen under considerable heat and pressure (Fig. **3**). Even humans are also capable of fixing nitrogen, such as manufacturing nitrogen fertilizer requires combining nitrogen with hydrogen in the presence of Iron as a catalyst, producing ammonia. Another case where nitric oxide is formed by combining nitrogen and oxygen under high temperature and pressure is in internal combustion engines of vehicles. Exhaustion of vehicles also emits large amounts of nitric oxide into the environment. Nitrogen fixation by anthropogenic ways is about 120 million tonnes per annum, of which 83% is from fertilizer manufacturing. Therefore, it is a significant component of the nitrogen cycle globally and can be compared with non-anthropogenic nitrogen fixation, which is about 170 million tonnes per annum [18].

Nitrification

It involves the process of conversion of ammonium into nitrate by nitrifying bacteria in the soil. Primarily, bacteria like *Nitrosomonas* species convert ammonia into nitrites by the oxidation of ammonium, and then bacteria like *Nitrobacter* converts nitrites into nitrates. Conversion of ammonia into nitrites or nitrates is important as ammonia gas is toxic for plants. Plants take these nitrogen compounds to form Nitrates and can enter groundwater due to their high solubility. Groundwater is a concern of drinking water, and increasing nitrate levels can cause methemoglobinemia or blue-baby syndrome in infants [19]. The nitrification process does not occur under acidic conditions; therefore, plants in acidic environments must be capable of utilizing ammonium as a nitrogen source.

Ammonification

It involves the process of conversion of organic nitrogen into ammonium. When a plant or animal dies, bacteria and fungi convert organic nitrogen back into ammonium within the remains. The organically bound nitrogen must be transformed into inorganic form as it is necessary to recycle its fixed nitrogen. In ammonification, the organic nitrogen of dead matter is converted to ammonia; the process requires hydrogen ions to form ammonium. The ammonium obtained is considered an appropriate nutrition source for many floral species. Certain species are incapable of utilizing ammonium effectively; therefore, in that case, nitrate acts as the source of nitrogen.

Denitrification

Denitrification is the final stage of the Nitrogen cycle. This process involves the reduction of nitrates back into nitrogen gas. Under anaerobic conditions, this process is performed by *Pseudomonas* and *Paracoccus* bacterial species. The denitrification rate is high wherever large concentrations of nitrate exist, as in agricultural lands that are flooded temporarily. Denitrification, in some cases, can be considered as a counter-balancing process to nitrogen fixation, as global rates of both processes are somewhat in a rough balance. Thus, no such change is observed over time-related to the total amount of nitrogen in the biosphere.

Anthropogenic activities alter the global nitrogen cycle, mostly in developed countries where industrial agriculture is at the highest level [20]. Due to industrialization and agricultural fertilization, Nitrous oxide has increased in the atmosphere. Nitrous oxide plays a deleterious role as a catalyst in atmospheric ozone. Nitrous oxide is the third greenhouse gas after carbon dioxide, that contributes to global warming [21].

Oxygen Cycle

The oxygen cycle describes the movement of oxygen through the atmosphere, lithosphere, and biosphere. In the atmosphere, 20.9% oxygen is available, which equates to 34×10^8 mol of oxygen [22]. The presence of oxygen in the atmosphere includes ozone (O_3), water vapor (H_2O), Carbon dioxide (CO_2), sulfur, and nitrogen oxides (SO_2, NO, N_2O). In the biosphere, 22% of oxygen is available; in the lithosphere, about 46.6% of oxygen is present mainly as silica minerals (SiO_2) and other oxide minerals. During photosynthesis, green plants release oxygen as a by-product in the atmosphere, which aerobic organisms use for Respiration. Aerobic organisms exhale carbon dioxide into the atmosphere and use plants during photosynthesis. This process maintains a balance of oxygen in the atmosphere. The presence of atmospheric oxygen has also contributed to the formation of ozone within the stratosphere as it absorbs noxious ultraviolet radiations.

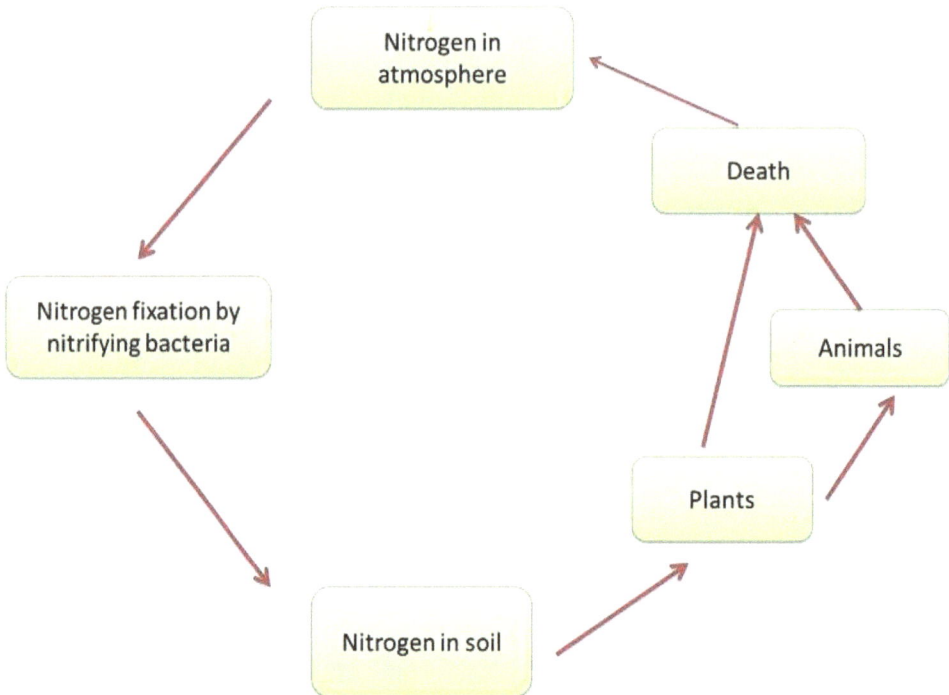

Fig. (3). Nitrogen cycle.

Nutrient cycling allows the flow of nutrients from the soil to consumers for the survival of organisms and then back again after death and decomposition. Nutrient cycling is a quality approach for environmental sustainability and net production to continue.

Sulfur Cycle

Sulphur is the main constituent of proteins, amino acids, and various biochemicals. Sulphur is significantly present in soil, and water and is abundantly reported in some minerals and rocks. It also occurs in certain compounds that induce air pollution, such as sulfur dioxide released by volcanic eruptions.

Mostly, sulphur dioxide emissions are related to anthropogenic activities. Another compound is hydrogen sulphide, which smells like rotten eggs and is released from deep-sea vents, volcanic eruptions, and habitats where decomposition of sulphur compounds occurs under anaerobic conditions. The global emissions of sulphur per annum are about 251 million tonnes. Various organically bound sulphur forms occur in soil and water environments. These include the proteins and sulphur compounds of dead organic matter. Cycling sulphur is a series of processes that transfer sulphur between rocks, rivers, and living systems. The global sulphur cycle includes sulphur species transformations through various oxidation states, which play an important role in geological and biological processes [23]. Microbes present in soil oxidize the organic sulphur to sulphate, which plants in their nutrition further utilize. Plants mostly absorb sulphate dissolved in the soil through roots. Anthropogenic activities have immensely altered the balance of sulphur cycle by toxicity, acid-mine drainage, and acid rain. However, in certain cases like agriculture, it is used for manufacturing fertilizers.

Nutrient Cycling in Higher Altitudes

The nutrient cycling process in higher altitudes varies greatly compared to the lowlands. The higher altitudes possess a unique ecosystem, thereby imposing a thrust on nutrient cycling. The reasons may include low temperature, fewer microbes, soil texture, and low enzymatic activity. The organic carbon content in the soil increases with altitude, but nutrient availability is decreased with elevation. The plant organic carbon is insensitive to the environment and thus remains constant, whereas there is a decrease in microbial biomass and soil carbon (Fig. **4**).

Further, organic matter degradation is also decreased due to the increased C:N ratio of litter. The availability of nitrogen in the soil decreases with soil depth at higher altitudes, and the same is observed in the case of phosphorus. The portion of phosphorus remaining in the soil is unavailable, as a reduction in phosphorus content at higher altitudes is due to large elevation and its storage in biomass [24]. The potassium content of the soil also differs with increasing altitude; high potassium on the soil surface is converted into soil solution [25]. In polar ice caps, nutrient levels are extremely low due to the immense decrease in temperature, less vegetation, and microbial and enzymatic activity. The presence of psychrophilic microbes in soils of Arctic regions has been reported to be efficiently involved in immobilization which eventually leads to less mineralization of N, P, and K [26]. The plant and animal diversity in higher altitudes have an influential role in nutrient cycling as their rate varies compared to low altitude areas.

Nutrient Cycling in Aquatic Ecosystems

Anthropogenic activities affect the nutrient cycles on land; they remove nutrients and discharge them into aquatic systems, resulting in eutrophication. Eutrophication changes the structure and functioning of the aquatic ecosystem. Eutrophication symptoms include an increase in benthic and epiphytic algae, phytoplankton, and bacterial biomass, forms algal bloom, anoxia, decreased water transparency, and increases in mortality incidence in fish and shellfish. Nutrient load responsible for eutrophication is mainly associated with C, N, and P. Excreta of aquatic animals also provide nutrients for primary producers, bacteria, and fungi. Some aquatic animals excrete nutrients in organic forms like Urea and inorganic form-ammonia and phosphate [13]. Nutrient cycling in the aquatic ecosystem involves:

Carbon Cycle

The carbon cycle describes the movement of carbon through the aquatic ecosystem. The carbon cycle is related to carbon dioxide, which is present in dissolved form in aquatic systems. Dissolved carbon dioxide stored in water is present either as carbonate or bicarbonate ions, preventing water from getting much acidic or basic. Carbon is also transported from terrestrial ecosystems due to weathering and soil erosion and later accumulated in sediments [27]. Increasing levels of carbon dioxide and other carbon-containing compounds in oceans causes an increase in acidity, about 30%, and changes the ocean chemistry (NOAA 2020). Aquatic plants use dissolved CO_2 and generate O_2 which other animals use for respiration. After the death of aquatic plants and plant-eating animals, it accumulates in the sediments as carbon sources and gets uplifted over time through weathering or geologic processes. Buried carbon in the sediments remains stored for millions of years in the earth's mantle to continue the carbon cycle [28]. Carbon dioxide is also present in marine water as a bicarbonate ion which is eventually fixed photosynthetically by algae and bacteria. Marine organisms also manufacture calcium carbonate shells by using bicarbonate ions and carbon dioxide.

Nitrogen Cycle

The nitrogen cycle in aquatic systems is similar to terrestrial systems, but the players are different. Nitrogen in water comes through runoff, precipitation, or directly from the atmosphere. N_2 cannot be used by phytoplankton; it needs nitrogen fixation, which is done by Cyanobacteria [28]. Primarily, the ammonification process is done in which organic nitrogen is converted into ammonia; then nitrification is done by Ammonia oxidizing bacteria *Nitrospira* and Thaumarchaeota that converts ammonia into nitrite and nitrate [29]. Nitrate is utilized by phytoplankton for growth and other processes. Finally, the denitrification process is done by aerobic bacteria-Thiosphaera pantotropha, *Bacillus subtilis* [30, 31], and other species of bacteria in which N2 is returned to the atmosphere.

The effects of microplastic biofilm on Nitrogen and Phosphorus cycling in freshwater systems were analyzed in a study. The results revealed that microplastic biofilms alter the P cycle by sorption and the N cycle by increasing denitrification in aquatic systems [32]. More research is needed to predict the influence of microplastic on biofilm formation and Nutrient cycling -P and N.

Phosphorus Cycle

Phosphorus is a limiting nutrient present in the rocks on earth. Due to weathering and erosions, it comes on the earth's surface, which terrestrial plants and animals then use. Though, phosphorus is required by living organisms in minor quantities as compared to carbon and nitrogen. It is considered a critical nutrient in different environments, particularly freshwater and terrestrial systems, due to its less availability. In the case of phosphorus cycling, a significant atmospheric phase is absent, which is observed in nitrogen and carbon cycles. The trace quantities of phosphorus compounds occur in particulates, resulting in low phosphorus input to the environment compared to the amount present in the soil. Due to runoff soil particles, fertilizer runoff from agricultural lands or industrial wastes is transported into aquatic systems mainly in the inorganic form- orthophosphate (PO_4^{3-}) and is converted into organic form by phytoplankton [33]. Aquatic animals utilize organic phosphorus, released back through excretion, and after the death of animals, phytoplankton gets settled down into sediments. By bacterial decomposition, organic phosphate is converted back into inorganic form and is uplifted through a geologic process.

Factors that Imbalance Nutrient Cycle

Environment unfriendly activities that affect the biogeochemical cycles and are responsible for climate change involve deforestation, fossil fuel burning, use of synthetic fertilizers, and Automobile emissions.

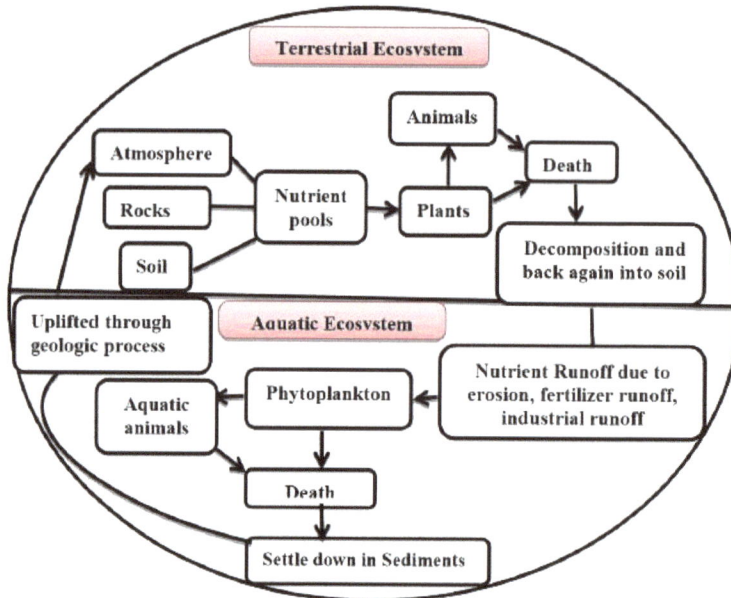

Fig. (4). Nutrient cycling.

Deforestation

The removal of the stand of trees from land that is then turned into a non-forest use is deforestation. Around 31% of the earth's surface is covered by forests [34]. Deforestation is a global environmental issue and a key contributor to climate change; it also causes a change in nutrient dynamics and Carbon sequestration [35]. About 11% of greenhouse gas emissions are caused due to deforestation. Rising demand for food and other trade products forces the conversion of natural vegetation into agricultural land, leading to soil erosion. According to the FAO, about 80% of deforestation is caused by agriculture. Leaching due to erosion reduces soil fertility and available nutrients pool, resulting in eutrophication in aquatic bodies. The soil heats up faster and reaches a higher temperature in deforested areas, leading to localized upward movements that increase cloud formation and generate more rainfall. Deforestation affects wind patterns, water vapor flows, and solar energy absorption, influencing local and global climate [36]. A study published in the journal Ecohydrology found higher soil and air temperatures in parts of the Amazon rainforest that were transformed into agricultural land, which can intensify drought conditions.

In contrast, forested land had evapotranspiration rates that were nearly three times higher, contributing more water vapor to the air [37]. Deforestation also produces CO_2 emissions [38] and contributes to global warming. Kelly and Jordan 2015 analyzed the effect of deforestation on soil and streams of the tropical Peninsula and found that available soil phosphorus reduces > 33% and 95% loss of dissolved inorganic nitrogen in the deforested catchment. 59% increase in stream discharge flow during the wet season shows deforested stream yields higher magnitude flood peaks.

Soil Erosion

The land is a resource for operations in agriculture. Unregulated land use increases the cause of soil erosion and loss of biodiversity. Intensive farming practices are reported as a leading cause of soil erosion and biodiversity loss. Soils are one of the main living areas of micro-organisms, which are involved in organic matter decomposition and soil nutrient cycling like nitrogen, carbon, phosphorus, and sulfur [39]. Erosion is commonly regarded as one of the primary soil threats [40]. Soil erosion is one of the key processes affecting nutrient cycling and eventually reducing soil fertility. Soil erosion decreases the number of available nutrients in the soil, resulting in severe modifications in the biological and chemical properties of the soil matrix. Worldwide, high erosion rates are persisting on more than 1.1 billion hectares of land [41]. The transportation of nutrient-rich soil from one place to another result in the redistribution of nutrients over the landscape. Some nutrients are lost to riverine systems throughout this process and eventually transported to marine systems. Soil erosion decreases the potential to isolate atmospheric carbon dioxide in soils, reducing primary productivity. Soil erosion is the significant way mineral-associated soil organic carbon is translocated in significant quantities into aquatic systems [42].

Fossil Fuel Burning

Fossil fuel is a natural process formed fuel due to the anaerobic decomposition of buried remains of plants and animals that liberate energy in combustion. High percentages of carbon are present in fossil fuels, including oil, coal, and natural gas; sometimes, peat is also considered a fossil fuel. In general, fossil fuels are categorized as non-renewable resources because they take millions of years to form, and known viable reserves are exhausted much faster than new ones [43]. Burning fossil fuels in industries results in additional nitrogen and nitrous oxide compounds in the atmosphere, dis-balances the natural nitrogen levels, and contributes to smog and acid rain. Excess nitrous oxide in the atmosphere is deposited back on land, which washes into nearby aquatic bodies and contributes to pollution, harmful algal blooms, and oxygen-deprived aquatic zones. Also, fossil fuel burning produces 30 billion tonnes of CO_2 per year [44], contributing to global warming and ocean acidification. Now, CNG (Compressed natural gas) replaces these fuels because they are cleaner and produce less toxic products.

Use of Synthetic Fertilizers

Man-made blends of chemicals and inorganic substances are synthetic fertilizers. They usually combine nitrogen, phosphorus, potassium, calcium, magnesium, and other elements in varying ratios. Unlike their organic counterparts, synthetic fertilizers instantly give the soil vital nutrients. At the end of the 19th century, synthetic fertilizers found their way in and paved the way for modern agricultural development. Their use increased crop yields and gave rise to an agricultural revolution that the world had never seen before. Synthetic fertilizers continue to have positive and negative effects. Synthetic fertilizers are added to harvest more yields in short timescales. Application of fertilizer to achieve food production and environmental stewardship should focus on "4R's"-right rate, place, time, and kind. The application of synthetic fertilizers alters the natural nutrient cycles. There are long-term harmful consequences of synthetic fertilizers. In the soil, synthetic fertilizers kill beneficial micro-organisms that convert dead remains of plants and humans into organic matter rich in nutrients. Synthetic fertilizers disrupt the natural soil composition. Plants growing in overly fertilized soil are deficient in Iron, zinc, carotene, vitamin C, copper, and protein. Plants harvested from such soils get converted into nitrites in the intestines upon consumption. These nitrites react with hemoglobin in the blood and cause methemoglobinemia, due to which suffocation occurs or death in severe cases.

A group of scholars [45] analyzed the impact of green fertilizers instead of synthetic fertilizers and/or animal manure on soil and crop production and found that application of bio-based mineral fertilizers induced small, statistically insignificant development in crop yield, soil fertility, and quality compared to synthetic fertilizers and animal manure, however, stimulates nutrient mobilization from the soil and increases the use efficiency of nutrients. For all reuse scenarios, economic and ecological avail were significantly higher than the reference.

A study examined the fertilizing properties and value of digestate (digested pulp) from three different agricultural biogas plants, and the results revealed that digestate was rich in nitrogen (up to 6.58 kg^{-1}) and potassium (up to 16.3 kg^{-1}) in the fresh matter. The value of digestate was estimated in the range of 6.08- 15.36 EUR Mg^{-1} [46]. The use of such waste by-products limits the use of synthetic fertilizers. Generating digestate also decreases the load of Chemical oxygen demand (COD) released to the environment.

Automobile Emissions

Automobiles are the major cause of global warming. Automobiles release Mono-nitrogen oxides (NO and NO_2), Carbon monoxide, and Carbon dioxide, which results in a change in nutrient cycling. Automobile emissions can be reduced by public and alternative transport, zoning towns, and other settlements. Another alternative for automobile emissions is biofuels, which do not directly contribute to greenhouse gas emissions. However, biofuels are problematic because the agricultural sector is fossil-fuel-intensive and often causes other environmental harm, such as soil depletion and oceanic dead zones.

Organic Matter

Organic matter is influential in determining the availability of nutrients to the plants. The organic compounds typically contain large stocks of nutrients in water, soil, and sediments. As the uptake of nutrients by plants is mostly in inorganic form thereby, organic matter decomposition is important to nutrient availability. Organic matter imparts energy for all microbial and faunal activities. This energy allows them to build the micro-aggregates, thereby authorizing soil hydraulic properties which further conserve organic matter. In the provision of the soil ecosystem, agricultural practices may have a substantial impact on the role of organic matter. In the case of the prohibition against burning sugarcane, crop residues in some countries may affect the carbon storage in soil. Organic matter stocks are also increased by pasture rotation practice which restores organic stocks by annual crops by inserting perennial grass lays, increasing organic matter stocks [47].

Importance of Nutrient Cycling in Plants and Humans

Soil supplies nutrients to plants and humans to complete their life cycle. For growth and reproduction, nutrient cycling is pre-eminent to meet metabolic demands. Nutrients such as Carbon, Oxygen, and Hydrogen are major elements that form the basic structures of living beings and are meant for Respiration. Another element that is required for structural roles is Calcium. Calcium plays an important role in the cell wall structure and maintains membrane permeability [48]. Once it has reached the cell, it is used to produce membrane voltage and regulate signal transduction systems and enzyme activation [49]. In plants, Signal transduction systems regulated by calcium ions include stress tolerance (Cold and dryness), touch stimulation response, and stomatal movements [50], while in humans, various genes control such systems.

Although another most valuable nutrient is magnesium, in plants, magnesium is a component of chlorophyll and functions in enzyme activation and ribosome stability [48]. In contrast, magnesium in humans is the fourth most abundant mineral, present about 25g in human bone [51], helps in blood pressure regulation, development of bones, maintaining normal heart rhythm, and helps in the transportation of Calcium across the membrane [52]. Phosphorus is an essential nutrient in the plant, and animal cells help plants in absorbing the sun's energy and transform it into growth and reproduction. Phosphorus in plants functions as electron carriers in chloroplasts and mitochondria; boosts root development, and improves the quality of crops [53], while Phosphorus in humans is an important constituent of cell membranes, DNA, and RNA and helps in the formation of bone, teeth, and ATP -Energy storage molecule [53]. The acute lethal dose of Phosphorus in humans is about 1 mg kg^{-1} [54]. For bone strength and muscle development, Phosphorus and other nutrients are needed. Nitrogen is a major component of genetic material DNA, by which genetic information is transmitted to upcoming generations of organisms, helps plants in growth and development, and is involved in plant processes- photosynthesis [55]. Nitrogen underlies the green color of plants and vegetated areas on earth due to N-containing chlorophyll proteins (enzymes). When plants are fed to animals, or when we consume plants, these plant proteins become animal proteins. Therefore, nitrogen is a key factor in the whole food system and interacts strongly with human management [56]. Nitrogen in humans helps in protein synthesis, proper digestion, and brain functioning [57]. Nitrogen is immensely important in the growth of a human fetus.

Nevertheless, for all living organisms, iron is also an important micronutrient because it plays a major role in metabolic processes such as respiration, photosynthesis, and DNA synthesis. In addition, Iron stimulates many metabolic pathways and is a prosthetic group integrant of many enzymes. Therefore, Iron is essential for a broad range of biological functions. The key causes of Iron Chlorosis are an imbalance between the solubility of Iron in the soil and the plant's demand for iron [58]. Iron assists plants in synthesizing chlorophyll and electron carriers and plays a role in activating some enzymes [48]. In humans, it is an essential component of hemoglobin (Around 70% of Iron in the human body is found in Red blood cells, 6% is a constituent of certain proteins, and 24% of such nutrient is stored as ferritin) and functions in the transport of oxygen, myelination of the spinal cord and cerebellar folds of the brain and is a cofactor of many enzymes [59]. Another essential nutrient that supports life is zinc. Zinc in plants plays a very important role in the metabolism by controlling carbonic anhydrase and hydrogenase activities, functions in the synthesis of hormones, and forms chloroplasts, auxin, and starch. Plants require zinc to regulate and maintain gene expression to tolerate environmental stress conditions [60]. While in humans, it is a constituent of many enzymes like Carbonic anhydrase, alkaline phosphatase, DNA, and RNA polymerase [61]. About 1.5-2.5g of zinc is found in the human body-60% found in muscles and 30% in bones [62]. Zinc also maintains the gut structure and gut immune functioning.

Thus, Nutrient cycling is important for living organisms' survival and better growth. Cycling nutrients is mandatory to maintain energy flow and link abiotic and biotic constituents within the ecosystem.

Effect of the Paucity of Nutrient Cycling in Plants and Humans

Nutrients play important roles in plants and humans for their normal life processes. Inadequate or excess of any one nutrient can cause hindrance and may result in deficiency symptoms [63]. Plant deficiency symptoms are shown in the general growth of stems, leaves, roots, and fruits. Nutrient deficiencies are influenced by soil conditions (pH), weather (drought or excessive rain), and farming practices (unsuitable rotations). Deficiency symptoms of nutrients in plants include:

Nitrogen: It causes stunted growth in plants because of reduced cell division, early maturity in crops which reduces yield and quality, leaves become yellow and die (Chlorosis) [48].

Phosphorus: Phosphorus deficiency symptoms in plants include: Leaves and stems turn dark green, Veins of leaves become purplish, and poor development of seeds and fruits. In certain situations, the tips of leaves will turn brown and die.

Potassium: Potassium deficiency occurs first in older leaves because potassium is very mobile in plants; weakening of roots and stems, reduced growth, brown scorching and curling of leaf tips, and yellowing between leaf veins are usual symptoms of potassium deficiency in plants. The undersides of the leaves can also have purple spots [64].

Magnesium: Magnesium forms the central atom of chlorophyll. In plants, magnesium is highly mobile, and signs of deficiency first appear on the lower leaves. The lower leaves have more severe symptoms because magnesium is transferred to new growth and includes interveinal Chlorosis and drooped leaves [65].

Boron: It affects vegetative and reproductive growth of plants, lateral branches of plants die, leaves become thick, and reduced fertility [66].

Molybdenum: It causes yellowing of leaves with marginal and interveinal Chlorosis, cupped leaves, and inhibition of nitrogen fixation in legumes, resulting in nitrogen-deficient plants [67].

Copper: Necrosis in younger leaves; in some cases, leaves may wilt, inhibits the growth of lateral branches, and make the color of flowers lighter than normal.

Zinc: It causes large losses in crop production and quality, Necrotic spots on leaves, young leaves are most affected, yellowing of leaves [68].

Iron: Its deficiency first occurs in younger leaves and causes interveinal Chlorosis, decreases vegetative growth, and causes yellow and green striping in grasses [69]. In comparison, the deficiency of carbon, oxygen, and hydrogen causes stunted growth in plants.

Nutrient Deficiency in Humans

The deficiency of nutrients in humans causes malnutrition, also known as hidden hunger. Financial and economic crises have a greater impact on nutrient deficiency, resulting in serious public health issues. Deficiency symptoms of various nutrients include:

Iron and Copper: Iron plays an important role in cellular respiration, and deficiency of such elements causes anemia, resulting in reduced growth, depressed immune function, and poor body temperature. While fatigue, constant vomiting, frail and fragile bones, problems with learning and memory, walking difficulties, increased cold sensitivity, light skin, prematurely grey hair, and loss of vision are typical signs and symptoms of copper deficiency.

Zinc: It is involved in cell metabolism and cell division. Its deficiency causes stunted growth, hair loss, delayed wound healing, delayed sexual maturation, eye and skin lesions, and taste alterations [70].

Magnesium: It is involved in protein synthesis, muscle function, and blood pressure control. Its deficiency causes vomiting, muscle cramps, and abnormal heart rhythms [71].

Molybdenum: Its deficiency causes lens dislocation, intellectual disability, and opisthosomas [72].

Phosphorus and Potassium: Phosphorus is present as phosphate in the blood serum, and its reduced concentration causes a disorder known as hypophosphatemia. The deficiency of such elements causes bone diseases such as Rickets in children and Osteomalacia in adults. Thus nutrients maintain physicochemical processes that are essential to life. Potassium deficiency normally happens when much fluid is lost from the body. Weakness, muscle cramps, heart palpitations, tingles and numbness, breathing problems, digestive system, and mood changes are typical signs and symptoms of potassium deficiency.

Furthermore, in terrestrial and aquatic systems, nutrient cycling and soil fertility play important roles which significantly impact society. Indirect and direct advantages to humans are derived from nutrient cycling. Millions of people earn from manufacturing essential commodities derived from terrestrial and aquatic systems, such as food. In rural areas of developing countries, livelihood is mainly dependent on harvesting food and non-food materials. The ecosystem and the processes within generate the non-commercialized products, which are further used for deriving benefits directly. Therefore, nutrient cycling and fertility significantly support producing and supplying various products for human consumption.

Nutrient Alterations

Alterations in nutrient cycling may result in nutrient excess or nutrient deficiency. Increased nutrients above the normal range may cause eutrophication of soils and aquatic bodies, whereas its reduction may lead to soil exhaustion and certain specific changes in marine systems. Nutrient excess due to fertilizer inputs and atmospheric deposits causes aquatic systems' eutrophication, including freshwater and marine. The eutrophication process refers to the fertilization of water bodies by previously scarce nutrients [73]. In the 1960s, nutrient alterations were observed in certain lakes, especially those flowing through industrial areas. This nutrient load was due to anthropogenic activities, including urban, agricultural, and industrial activities. These sources' runoff mainly comprised carbon, nitrogen, and phosphorus [74]. Eutrophication results in the structural and functional changes of aquatic systems, thereby altering the output. The signs indicating eutrophication include oxygen depletion, reduced water transparency, water taste, odor, and treatment issues.

Furthermore, it also indicates increasing bacterial biomass, phytoplankton, and epiphytic algae. It surely affects biodiversity by increasing fish and shellfish mortality and coral mortality in coastal areas. The increased growth of phytoplankton enhances nitrogen levels, lowering oxygen in marine systems. The remedial strategies are costly and generally include reducing nutrient input up to tolerable levels.

Nutrient deficiency is mainly observed in agricultural lands of developing countries. Removal of nutrient inputs from anthropogenic or natural sources results in declining soil fertility. Mostly nutrient-deficient agricultural lands possess soil derived from basement rocks. The phosphorus content of this soil is inherently low, resulting in slow nitrogen fixation. The soil texture is sandy and possesses low organic carbon; therefore, it loses nitrogen through leaching. Thus, when farmers perceive that the expected yield will not be achieved due to nutrient deficiency, they hesitate to invest in fertilizers.

Similarly, a reduction in water flow from terrestrial systems into aquatic systems may also result in similar conditions. Alterations in nutrient cycling and the mechanisms involved mainly operate at much smaller scales. Certain efforts are mandatory to enhance the quality of nutrient cycling from inputs to the needs of agricultural activity. Thereby limiting the risk of nutrient leaching from terrestrial systems to freshwater and eventually to marine systems.

Nutrient Management

The processes involved in nutrient cycling are themselves the key to nutrient management. In terrestrial and aquatic systems, nutrients move from one compartment to another in one form or another. Further, there is a continuous input of nutrients from the atmosphere. During the process, most of the nutrients are conserved. However, there are atmospheric inputs and losses through denitrification, erosion, and ammonia volatilization [75]. While soil erosion is reducing agricultural productivity, it is also impacting the aquatic bodies worldwide. The alterations in nutrient cycling are also increasing due to anthropogenic activities. These activities led to the increased concentrations of carbon dioxide (CO_2), nitric oxide (NO), methane (CH_4), and nitrous oxide (N_2O), hence contributing to global warming [75]. Thereby incorporation of nutrient management strategies is mandatory for balance. The management plans must reduce soil erosion, increase nutrient use efficiency, conserve air water and soil quality and monitor macro and micro-nutrient cycling. The primary source of carbon in soil organic matter is plant-derived materials. Carbon is the most abundant constituent, the compost, manure, and other organic sources are helpful for organic carbon and other nutrient cycling. Organic carbon can also increase the availability of essential micro-nutrients by contributing to the formation of chelate compounds. The management strategies in increasing carbon inputs to the environment may reduce soil erosion and improve soil quality factors, including water holding capacity, porosity, cation exchange capacity, aggregate formation, and drainage [76].

In aquatic systems, alterations in nutrient cycling due to anthropogenic activities have increased with time. It has been reported that the limiting nutrient in the case of freshwaters is phosphorus (P), whereas, in marine systems, losses of nitrogen (N) through denitrification are observed [77]. In a study on comparative denitrification of marine, brackish, and freshwater, no large differences in the rates of denitrification between freshwater and marine systems were observed [78]. However, denitrification is an important loss for the marine system, as nitrogen fixation is much greater in freshwater than in marine systems [79]. The difference in the phosphorus cycle between freshwater and marine systems exists as annually; Phosphorus is re-mineralized from sediments and returned to the overlying water in the case of the marine system [80]. Phosphorus release from the sediments depends highly on sulfate concentrations, which can be considered a substitute for salinity [81]. Hence, phosphorus loss is preferential and a limiting nutrient in freshwater systems. Human activities greatly affect nutrient cycling, roughly doubling nitrogen flux and tripling phosphorus concentrations [82]. Therefore, management strategies are required to maintain all the nutrients, as managing one nutrient may be problematic. The combined voluntary and mandatory approaches for nutrient management are the best solutions. Reducing anthropogenic activities, including leaching from agricultural sources, manure processing, fossil fuel emissions, and few non-point inputs, is immensely required to balance nutrient cycling. Complementary to this, the scarcity of nutrients may be overwhelmed by enhancing sinks for the nutrients, for example, wetlands, riparian zones, and ponds which are considered significant sinks for nutrients, especially nitrogen and phosphorus. The quality of the environment depends on the availability of nutrients [83]. Nutrient balancing depends on accounting for the approaches related to inputs, stores, and outputs of nutrients. Focusing on balancing nutrient cycling and its management strategies is mandatory for enhancing the sustainability of our environment.

CONCLUSION

Nutrients on earth are available in limited quantities. To grow and reproduce, living organisms need food. Any food that an organism needs to survive, grow or reproduce is called a nutrient. Nutrients move within and between ecosystems for the sustenance of life, referred to as Nutrient cycling. A unique cycle follows every nutrient, such as carbon, oxygen, phosphorus, magnesium, *etc*. Some elements, such as oxygen and nitrogen, cycle rapidly and are readily available for organisms. Others, such as phosphorus, magnesium, *etc*., require time to cycle because they are slowly released. Typically, such slow-cycling nutrients become the limiting factors for the growth of plants. For this purpose, synthetic fertilizers supply certain nutrients to crop species. In nutrient cycling, various nutrients for life move from the respective nutrient pool (abiotic) to biotic components and back again. Nutrient cycling links the abiotic and biotic components through the flow of nutrients. Thus, nutrient cycling is one of the important ecosystem services that maintain the net balance of available nutrients. Anthropogenic activities altering the nutrient cycle balance may dramatically affect the interactions among different ecosystems. Moreover, growing changes in the climate of the earth are significantly affecting nutrient cycling.

CONSENT FOR PUBLICATION

Not applicable.

CONFLICT OF INTEREST

The author declares no conflict of interest, financial or otherwise.

ACKNOWLEDGEMENTS

The author acknowledges Lovely Professional University, Punjab, for providing the necessary facilities and support in formulating the manuscript.

REFERENCES

[1]   Bohn HL, Myer RA, O'Connor GA. Soil chemistry. John Wiley & Sons 2002.

[2]   Abou el Magd MM, Hoda MA. Relationships, growth, yield of broccoli with increasing N, P or K ratio in a mixture of NPK fertilizers. Annals Agric Sci Moshtohor> 2005; 43(2): 791-805.

[3]   Arancon NQ, Edwards CA, Bierman P, Metzger JD, Lucht C. Effects of vermicomposts produced from cattle manure, food waste and paper waste on the growth and yield of peppers in the field. Pedobiologia (Jena) 2005; 49(4): 297-306.
[http://dx.doi.org/10.1016/j.pedobi.2005.02.001]

[4]   Wardle DA, Bardgett RD, Klironomos JN, Setälä H, van der Putten WH, Wall DH. Ecological linkages between aboveground and belowground biota. Science 2004; 304(5677): 1629-33.
[http://dx.doi.org/10.1126/science.1094875] [PMID: 15192218]

[5]   Moyano FE, Manzoni S, Chenu C. Responses of soil heterotrophic respiration to moisture availability: An exploration of processes and models. Soil Biol Biochem 2013; 59: 72-85.
[http://dx.doi.org/10.1016/j.soilbio.2013.01.002]

[6]   Fischer C, Leimer S, Roscher C, *et al.* Plant species richness and functional groups have different effects on soil water content in a decade long grassland experiment. J Ecol 2019; 107(1): 127-41.
[http://dx.doi.org/10.1111/1365-2745.13046]

[7]   Lavelle P, Dugdale R, Scholes R, *et al.* Nutrient cycling. Ecosystems and Human Well-Being: Current State and Trends: Findings of the Condition and Trends Working Group, Island Press, Washington, Covelo, London. 2005 Dec 14.Bormann FH, Likens GE. Nutrient cycling. Science 1967; 155(3761): 424-9.
[PMID: 17737551]

[8]   Bormann FH, Likens GE. Nutrient Cycling. Science 1967; 155(3761): 424-9.
[http://dx.doi.org/10.1126/science.155.3761.424] [PMID: 17737551]

[9]   Ponge JF. The soil as an ecosystem. Biol Fertil Soils 2015; 51(6): 645-8.
[http://dx.doi.org/10.1007/s00374-015-1016-1]

[10]   Fujii K. Soil acidification and adaptations of plants and microorganisms in Bornean tropical forests. Ecol Res 2014; 29(3): 371-81.
[http://dx.doi.org/10.1007/s11284-014-1144-3]

[11]   Ren X, Zeng G, Tang L, *et al.* Sorption, transport and biodegradation – An insight into bioavailability of persistent organic pollutants in soil. Sci Total Environ 2018; 610-611: 1154-63.
[http://dx.doi.org/10.1016/j.scitotenv.2017.08.089] [PMID: 28847136]

[12]   Kauppi P, Kämäri J, Posch M, Kauppi L, Matzner E. Acidification of forest soils: Model development and application for analyzing impacts of acidic deposition in Europe. Ecol Modell 1986; 33(2-4): 231-53.
[http://dx.doi.org/10.1016/0304-3800(86)90042-6]

[13]   Sterner RW, Elser JJ. Ecological stoichiometry: the biology of elements from molecules to the biosphere. Princeton university press 2002.

[14]   Filippelli GM. The global phosphorus cycle. Rev Mineral Geochem 2002; 48(1): 391-425.
[http://dx.doi.org/10.2138/rmg.2002.48.10]

[15]   Turner BL, Frossard E, Baldwin DS, Eds. Organic Phosphorus in the environment. CABI Pub.; 2005.Carroll SB, Salt SD. Ecology for gardeners. Timber Press 2004.

[16]   Carroll SB, Salt SD. Ecology for gardeners. Timber Press 2004.

[17]   Smith BE, Richards RL, Newton WE, Eds. Catalysts for nitrogen fixation: nitrogenases, relevant chemical models and commercial processes. Springer Science & Business Media 2013.

[18]   Freedman B. Flows and cycles of nutrients. Environ Sci (Ruse) 2018.

[19]   Vitousek PM, Aber JD, Howarth RW, *et al.* Human alteration of the global nitrogen cycle: sources and consequences. Ecol Appl 1997; 7(3): 737-50.
[http://dx.doi.org/10.1890/1051-0761(1997)007[0737:HAOTGN]2.0.CO;2]

[20]   Holland EA, Dentener FJ, Braswell BH, Sulzman JM. Contemporary and pre-industrial global reactive nitrogen budgets. Biogeochemistry 1999; 46(1-3): 7-43.
[http://dx.doi.org/10.1007/BF01007572]

[21]   Howarth RW, Bringezu S, Bekunda M, *et al.* Rapid assessment on biofuels and environment: overview and key findings. Biofuels: environmental consequences and interactions with changing land use. 2009:1-3.Petsch ST. The global oxygen cycle. Treatise on geochemistry 2009.

[22]   Petsch ST. The global oxygen cycle. Treatise on geochemistry 2003.
[http://dx.doi.org/10.1016/B0-08-043751-6/08159-7]

[23]   Madigan MT, Martinko JM, Parker J. Brock biology of micro-organisms 1997.

[24]   Wu Y, Zhou J, Yu D, *et al.* Phosphorus biogeochemical cycle research in mountainous ecosystems. J Mt Sci 2013; 10(1): 43-53.
[http://dx.doi.org/10.1007/s11629-013-2386-1]

[25]   Dangwal LR, Singh T, Singh A, Sharma A. Plant diversity assessment in relation to disturbances in subtropical chirpine forest of the western Himalaya of district Rajouri, J&K, India. International Journal of Plant. Animal and Environmenal Sciences 2012; 2(2): 206-13.

[26]   Jonasson S, Michelsen A, Schmidt IK, Nielsen EV, Callaghan TV. Microbial biomass C, N and P in two arctic soils and responses to addition of NPK fertilizer and sugar: implications for plant nutrient uptake. Oecologia 1996; 106(4): 507-15.
[http://dx.doi.org/10.1007/BF00329709] [PMID: 28307451]

[27]   Forsberg C. Importance of sediments in understanding nutrient cyclings in lakes. Hydrobiologia 1989; 176-177(1): 263-77.
[http://dx.doi.org/10.1007/BF00026561]

[28]   Ducklow H, Steinberg D, Buesseler K. Upper ocean carbon export and the biological pump. Oceanography (Wash DC) 2001; 14(4): 50-8.
[http://dx.doi.org/10.5670/oceanog.2001.06]

[29]   Hatzenpichler R. Diversity, physiology, and niche differentiation of ammonia-oxidizing archaea. Appl Environ Microbiol 2012; 78(21): 7501-10.
[http://dx.doi.org/10.1128/AEM.01960-12] [PMID: 22923400]

[30]   Robertson LA, Kuenen JG. Thiosphaera pantotropha gen. nov. sp. nov., a Facultatively Anaerobic,

Facultatively Autotrophic Sulphur Bacterium. Microbiology (Reading) 1983; 129(9): 2847-55.
[http://dx.doi.org/10.1099/00221287-129-9-2847]

[31]   Yang XP, Wang SM, Zhang DW, Zhou LX. Isolation and nitrogen removal characteristics of an aerobic heterotrophic nitrifying–denitrifying bacterium, *Bacillus subtilis* A1. Bioresour Technol 2011; 102(2): 854-62.
[http://dx.doi.org/10.1016/j.biortech.2010.09.007] [PMID: 20875733]

[32]   Chen X, Chen X, Zhao Y, Zhou H, Xiong X, Wu C. Effects of microplastic biofilms on nutrient cycling in simulated freshwater systems. Sci Total Environ 2020; 719: 137276.
[http://dx.doi.org/10.1016/j.scitotenv.2020.137276] [PMID: 32114222]

[33]   Ruttenberg KC. 1013—The global phosphorus cycle 2014; 499-558.

[34]   Bradford A. Deforestation: facts, causes & effects. Life Science 2015.

[35]   Williams M. A new look at global forest histories of land clearing. Annu Rev Environ Resour 2008; 33(1): 345-67.
[http://dx.doi.org/10.1146/annurev.environ.33.040307.093859]

[36]   Khanna J, Medvigy D, Fueglistaler S, Walko R. Regional dry-season climate changes due to three decades of Amazonian deforestation. Nat Clim Chang 2017; 7(3): 200-4.
[http://dx.doi.org/10.1038/nclimate3226]

[37]   Oliveira G, Brunsell NA, Moraes EC, *et al.* Effects of land-cover changes on the partitioning of surface energy and water fluxes in AMAZONIA using high-resolution satellite imagery. Ecohydrology 2019; 12(6): e2126.
[http://dx.doi.org/10.1002/eco.2126]

[38]   Longobardi P, Montenegro A, Beltrami H, Eby M. Deforestation induced climate change: effects of spatial scale. PLoS One 2016; 11(4): e0153357.
[http://dx.doi.org/10.1371/journal.pone.0153357] [PMID: 27100667]

[39]   Liu G, Jin M, Cai C, Ma C, Chen Z, Gao L. Soil microbial community structure and physicochemical properties in amomumtsaoko-based agroforestry systems in the Gaoligong Mountains, Southwest China. Sustainability (Basel) 2019; 11(2): 546.
[http://dx.doi.org/10.3390/su11020546]

[40]   Orgiazzi A, Panagos P. Soil biodiversity and soil erosion: It is time to get married. Glob Ecol Biogeogr 2018; 27(10): 1155-67.
[http://dx.doi.org/10.1111/geb.12782]

[41]   Berc J, Lawford R, Bruce J, Mearns L, Easterling D. Conservation Implications of Climate Change: Soil Erosion and Runoff from Croplands: A Report from the Soil and Water Conservation Society 2003.

[42]   Mandal D, Ngachan SV. Role of soil erosion and deposition in stabilization and destabilization of soil organic carbon. Carbon Management in Agriculture for mitigating greenhouse effect 2012.

[43]   Starr GC, Lal R, Malone R, Hothem D, Owens L, Kimble J. Modeling soil carbon transported by water erosion processes. Land Degrad Dev 2000; 11(1): 83-91.
[http://dx.doi.org/10.1002/(SICI)1099-145X(200001/02)11:1<83::AID-LDR370>3.0.CO;2-W]

[44]   Miller G, Spoolman S. Environmental science: problems, connections and solutions. Cengage Learning 2007.

[45]   Ambrose J. Carbon emissions from fossil fuels could fall by 2.5 bn tonnes in 2020. 2020.

[46]   Vaneeckhaute C, Meers E, Michels E, Ghekiere G, Accoe F, Tack FMG. Closing the nutrient cycle by using bio-digestion waste derivatives as synthetic fertilizer substitutes: A field experiment. Biomass Bioenergy 2013; 55: 175-89.
[http://dx.doi.org/10.1016/j.biombioe.2013.01.032]

[47]   Czekała W, Lewicki A, Pochwatka P, *et al.* Digestate management in polish farms as an element of the

nutrient cycle. J Clean Prod 2020; 242: 118454.
[http://dx.doi.org/10.1016/j.jclepro.2019.118454]

[48]   Franzluebbers AJ, Stuedemann JA, Schomberg HH, Wilkinson SR. Soil organic C and N pools under long-term pasture management in the Southern Piedmont USA. Soil Biol Biochem 2000; 32(4): 469-78.
[http://dx.doi.org/10.1016/S0038-0717(99)00176-5]

[49]   Soetan KO, Olaiya CO, Oyewole OE. The importance of mineral elements for humans, domestic animals and plants-A review. Afr J Food Sci 2010; 4(5): 200-22.

[50]   Brini M, Ottolini D, Calì T, Carafoli E. Calcium in health and disease. Interrelations between essential metal ions and human diseases. Springer 2013; pp. 81-137.
[http://dx.doi.org/10.1007/978-94-007-7500-8_4]

[51]   Allen GJ, Chu SP, Harrington CL, *et al.* A defined range of guard cell calcium oscillation parameters encodes stomatal movements. Nature 2001; 411(6841): 1053-7.
[http://dx.doi.org/10.1038/35082575] [PMID: 11429606]

[52]   Joy EJM, Young SD, Black CR, Ander EL, Watts MJ, Broadley MR. Risk of dietary magnesium deficiency is low in most African countries based on food supply data. Plant Soil 2013; 368(1-2): 129-37.
[http://dx.doi.org/10.1007/s11104-012-1388-z]

[53]   Garrett RD, Rausch LL. Green for gold: social and ecological tradeoffs influencing the sustainability of the Brazilian soy industry. J Peasant Stud 2016; 43(2): 461-93.
[http://dx.doi.org/10.1080/03066150.2015.1010077]

[54]   Metson GS, Bennett EM, Elser JJ. The role of diet in phosphorus demand. Environ Res Lett 2012; 7(4): 044043.
[http://dx.doi.org/10.1088/1748-9326/7/4/044043]

[55]   Robles Á, Aguado D, Barat R, *et al.* New frontiers from removal to recycling of nitrogen and phosphorus from wastewater in the Circular Economy. Bioresour Technol 2020; 300: 122673.
[http://dx.doi.org/10.1016/j.biortech.2019.122673] [PMID: 31948770]

[56]   Leghari SJ, Wahocho NA, Laghari GM. HafeezLaghari A, MustafaBhabhan G, HussainTalpur K, Bhutto TA, Wahocho SA, Lashari AA. Role of nitrogen for plant growth and development: A review. Adv Environ Biol 2016; 10(9): 209-19.

[57]   Galloway JN. The global nitrogen cycle: changes and consequences. Environ Pollut 1998; 102(1): 15-24.
[http://dx.doi.org/10.1016/S0269-7491(98)80010-9]

[58]   Hanrahan G, Chan G. Nitrogen encyclopedia of analytical science, 2005.

[59]   Rout GR, Sahoo S. Role of Iron in plant growth and metabolism. Reviews in Agricultural Science 2015; 3: 1-24.
[http://dx.doi.org/10.7831/ras.3.1]

[60]   Malhotra VK. Biochemistry for Students. 10th ed., New Delhi, India: Jaypee Brothers Medical Publishers Ltd 1998.

[61]   Cakmak I. Plant nutrition research: Priorities to meet human needs for food in sustainable ways. Plant Soil 2002; 247(1): 3-24.
[http://dx.doi.org/10.1023/A:1021194511492]

[62]   Arinola OG. Essential trace elements and metal binding proteins in Nigerian consumers of alcoholic beverages. Pak J Nutr 2008; 7(6): 763-5.
[http://dx.doi.org/10.3923/pjn.2008.763.765]

[63]   Sloup V, Jankovská I, Nechybová S, Peřinková P, Langrová I. Zinc in the animal organism: a review. Sci Agric Bohem 2017; 48(1): 13-21.

[http://dx.doi.org/10.1515/sab-2017-0003]

[64]   Underwood EJ. Trace elements in human and animal nutrition 3rd ed., 1971.

[65]   Datnoff LE, Elmer WH, Huber DM. Mineral nutrition and plant disease. American Phytopathological Society 2007.

[66]   Hermans C, Vuylsteke M, Coppens F, *et al.* Systems analysis of the responses to long-term magnesium deficiency and restoration in *Arabidopsis thaliana.* New Phytol 2010; 187(1): 132-44.
        [http://dx.doi.org/10.1111/j.1469-8137.2010.03257.x] [PMID: 20412444]

[67]   Koshiba T, Kobayashi M, Matoh T. Boron nutrition of tobacco BY-2 cells. V. oxidative damage is the major cause of cell death induced by boron deprivation. Plant Cell Physiol 2009; 50(1): 26-36.
        [PMID: 19054807]

[68]   Sardesai VM. Molybdenum: an essential trace element. Nutr Clin Pract 1993; 8(6): 277-81.
        [http://dx.doi.org/10.1177/0115426593008006277] [PMID: 8302261]

[69]   Alloway BJ. Zinc in Soils and Crop Nutrition. International fertilizer Industry Association and International Zinc Association. Brussels, Belgium and Paris 2004.

[70]   Abadía J, López-Millán AF, Rombolà A, Abadía A. Organic acids and Fe deficiency: a review. Plant Soil 2002; 241(1): 75-86.
        [http://dx.doi.org/10.1023/A:1016093317898]

[71]   Ikeda M, Ikui A, Komiyama A, Kobayashi D, Tanaka M. Causative factors of taste disorders in the elderly, and therapeutic effects of zinc. J Laryngol Otol 2008; 122(2): 155-60.
        [http://dx.doi.org/10.1017/S0022215107008833] [PMID: 17592661]

[72]   Whang R, Hampton EM, Whang DD. Magnesium homeostasis and clinical disorders of magnesium deficiency. Ann Pharmacother 1994; 28(2): 220-6.
        [http://dx.doi.org/10.1177/106002809402800213] [PMID: 8173141]

[73]   Mendel RR, Kruse T. Cell biology of molybdenum in plants and humans. BiochimicaetBiophysicaActa (BBA)-. Molecular Cell Research 2012; 1823(9): 1568-79.

[74]   Carpenter EJ, Montoya JP, Burns J, Mulholland MR, Subramaniam A, Capone DG. Extensive bloom of a N2-fixing diatom/cyanobacterial association in the tropical Atlantic Ocean. Mar Ecol Prog Ser 1999; 185: 273-83.

[75]   Howarth RW, Anderson DB, Cloern JE, *et al.* Nutrient pollution of coastal rivers, bays, and seas. Issues in ecology 2000; (7): 1-6.

[76]   Liu S, Zamanian K, Schleuss PM, Zarebanadkouki M, Kuzyakov Y. Degradation of Tibetan grasslands: Consequences for carbon and nutrient cycles. Agric Ecosyst Environ 2018; 252: 93-104.
        [http://dx.doi.org/10.1016/j.agee.2017.10.011]

[77]   Delgado JA, Follett RF. Carbon and nutrient cycles. J Soil Water Conserv 2002; 57(6): 455-64.

[78]   Conley DJ. Biogeochemical nutrient cycles and nutrient management strategies InMan and River Systems 1999; 87-96.

[79]   Seitzinger SP. Denitrification in freshwater and coastal marine ecosystems: Ecological and geochemical significance. Limnol Oceanogr 1988; 33(4part2): 702-24.
        [http://dx.doi.org/10.4319/lo.1988.33.4part2.0702]

[80]   Howarth RW, Marino R, Lane J, Cole JJ. Nitrogen fixation in freshwater, estuarine, and marine ecosystems. 1. Rates and importance. Limnol Oceanogr 1988; 33(4_part_2): 669-87.
        [http://dx.doi.org/10.4319/lo.1988.33.4_part_2.0669]

[81]   Caraco N, Cole J, Likens G. A comparison of phosphorus immobilization in sediments of freshwater and coastal marine systems. Biogeochemistry 1990; 9(3): 277-90.
        [http://dx.doi.org/10.1007/BF00000602]

[82]   Caraco NF, Cole JJ, Likens GE. Evidence for sulphate-controlled phosphorus release from sediments of aquatic systems. Nature 1989; 341(6240): 316-8.
[http://dx.doi.org/10.1038/341316a0]

[83]   Lee GF, Jones RA, Rast W. Availability of Phosphorus to phytoplankton and its implications for phosphorus management strategies. Phosphorus management strategies for lakes 1980.

Environmental Microbiology: Advanced Research, 2022, 105-136

CHAPTER 6

Microbial Biosensors for Environmental Monitoring

Ritu Bala¹, Manpreet Kaur Somal², Mukesh Kumar¹, Arun Karnwal¹ and **Rohan Samir Kumar Sachan¹,***

¹ Department of Microbiology, School of Bioengineering and Biosciences, Lovely Professional University, Phagwara-144411, Punjab, India

² Department of Biotechnology, School of Bioengineering and Biosciences, Lovely Professional University, Phagwara-144411, Punjab, India

Abstract: Unchecked disposal of substances or compounds such as organic/inorganic heavy metals, polychlorinated biphenyls (PCBs), herbicides, pesticides, phenolic and nitrogenous compounds, and polycyclic aromatic hydrocarbons (PAHs) ubiquitously present in the environment poses a global concern. This requires constant monitoring of environmental pollutants. Biological-based monitors and biosensors with high specificity and sensitivity are applied to monitor and check the level of pollutants. These are biological-based methods used for the intervention of environmental pollutants as analytes. The widely used biosensors are made by immobilizing various enzymes, antibodies, whole cells in the devices, and transducers. Microbial biosensor devices sense the substances in the environment through the various biochemical reactions of the microorganisms incorporated in the devices. However, with the ease of genetic modification techniques like genetic engineering technologies, various microorganisms have gained immense popularity as ideal candidates for developing biosensors. The microbial biosensors' inexpensiveness, compactness, and portability offer advantages over conventional chemical sensors. The most significant aspect of microbial biosensors is the *in situ* detection capability, and real-time analysis has enhanced their acceptability and applicability in environmental monitoring. The following chapter deals with microbial biosensors to detect air, water, and soil pollutants.

Keywords: Analysis, Biosensors, Environment, Genetic Engineering, Microbes, Monitoring, Pollutants, Portability.

INTRODUCTION

Reporting and analysis of the presence of toxins, pollutants and pathogens that negatively impact the environment and human society are needed continuously on

---

* **Corresponding author Rohan Samir Kumar Sachan:** Department of Microbiology, School of Bioengineering and Biosciences, Lovely Professional University, Phagwara-144411, Punjab, India; E-mails: sachan.rohan@yahoo.com

a primary basis [1]. Conventional chromatography methods, *i.e.*, high-performance liquid chromatography (HPLC) and gas chromatography, are bulky and slow approaches used for the issues mentioned above worldwide. Henceforth, biosensors are gaining more attention than chromatography methods because of their rapidness and ability to monitor on-site trace levels of targets [2]. Biosensors are devices with biological molecules and physical transducers to provide a signal output that can give a measurable signal by converting the signal response with analytes [3]. The detectable signal in response to the analyte is directly proportional to the molar concentration of the analyte and also provides analytics for semi-quantitative and quantitative data. A whole-cell biosensor uses microorganisms consisting of various enzymes as bio-sensing elements, referred to as microbial biosensors [4]. The enzymes produced by the microorganisms show a high degree of specificity towards selected analytes without having the necessity of cost purification and time consumption, as well as neglecting minor adverse effects of the operating environment [5]. The immobilization process between the elements and the transducers should be stable and smooth to transfer the signals and responses to the transducers from the recognition elements. The necessity for achieving a reliable and standardized microbial biosensor is met through the integration of the microorganisms onto the transducers [6]. The signal quality, transferred to the transducer from the microorganisms, and the reuse of the biosensor, are determined by immobilization, which plays an important role in the development of the whole-cell biosensors (Fig. **1**).

Various techniques like adsorption, entrapment, encapsulation, or cross-linking that are some of the conventional immobilization methods, are utilized, and all these techniques either suffer from the negative effects under bad conditions or show poor long-term stability. For effective immobilization, nanotechnology offers advancements for a better alternative by using nanoparticles, nanotubes, and fibre optics which can provide the stability of elements and promote higher reliability [7]. Another crucial part of the microbial biosensor is the transducer which tends to convert the response from microbes into a measurable signal [8]. Currently, another technique for microbial biosensors has been proposed, which relies on optical transducers as the primary transducer used in the past decade, what we commonly know as microbial fuel cells (MFC). From the biodegradable organic compounds, sustainable electricity can be generated through microbial metabolism, in which MFCs provide selective sensing capability and high sensitivity [9].

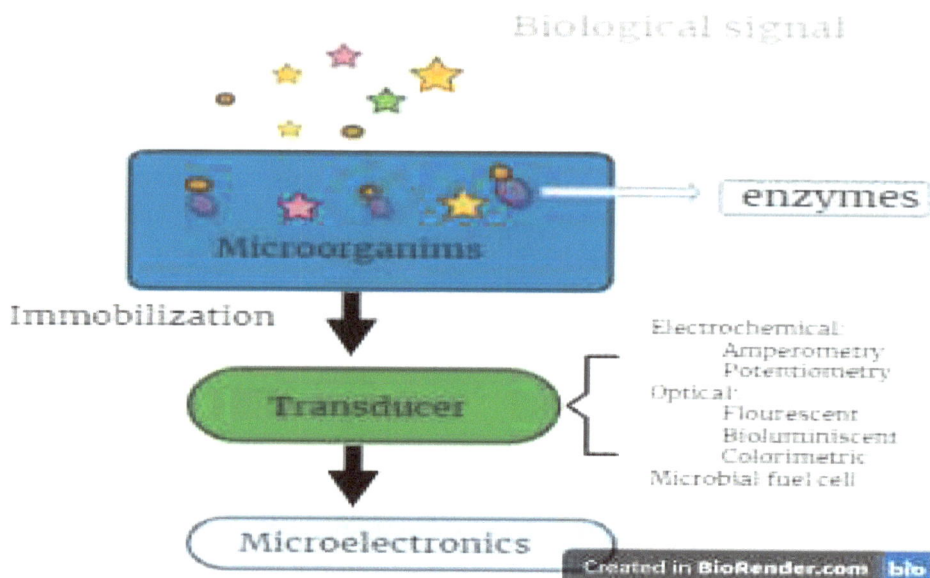

Fig. (1). A general representation of a whole-cell biosensor.

The principle behind applying microbial biosensors is to include the microbial respiration inhibition by the analyte of interest, such as environmental pollutants. For instance, microbial-based biosensors have been used to detect toxicity and genotoxicity to monitor water quality. However, substrate entry and products formed through the cell wall provide a slow signal. So, it decreases the biosensors' efficiency compared to the enzymes- based whole-cell biosensors [4]. Favourably, various factors or stimulants like physical (freezing and thawing), chemical (organic solvents and detergents), and enzymatic methods (papain, lysozyme) can increase the permeability of the cells [10]. Organic solvents are primarily used in different techniques which create minute pores in the cell membrane by removing some lipids which permeate more diffusion of small analytes inside the cell. BOD-based biosensors have a major significance in the coming future where polymers like protein, starch, lipid, *etc.,* are metabolized after being broken down into monomers [11]. The exo-expression or production of cellulase enzyme on the cell's surface has been genetically engineered. The whole cell has been shown to hydrolyze the cellulose incorporated media and use air-induced breakdown polymers before biosensor analysis can be replaced.

The viable cell-based biosensors are another approach to blocking unwanted transport systems and metabolic pathways [12]. Recombinant DNA technology can help produce enzyme-rich whole cells and engineer a limited number of enzymes that can catalyze side effects. Light emission-based microbial biosensors like luminescent bacteria are used as sensitive elements that could provide rapid detection. Such bacteria are non-invasive in many biological systems commonly found in marine (Vibro fischeri) and terrestrial (Photorhabdus luminescens) environments. Genetically engineered microorganisms are new trends used *in-vitro* to develop the bioluminescent-based whole-cell biosensor for detecting heavy metal, organic, and pesticide contamination [13]. Biosensors are typically developed with the help of microorganisms constructed using a plasmid. The genes are incorporated to code a gene sequence for the luciferase enzyme that carries out detection once placed under control conditions to recognize the analyte of interest [14].

Various microbial or whole-cell biosensors applications have been used widely in different areas such as environmental pollution monitoring, clinical diagnostics, food, and fermentation industries. This is due to inherent facts like cost-effective, stable, and fastidious response, and the biosensors are rapid, simple, and portable. Microbial biosensors are an effective screening method for environmental monitoring to identify and detect pollutants like organic and inorganic wastes, which can be dangerous to human or animal health [15]. The affordable price and reliability of the procedural methods are required to ensure the process is in control and maintain the quality of products in the food and fermentation industries [16]. Also, in clinical diagnostics, the detection rate is fast and accurate, and such inexpensive devices are urgently required to monitor clinically essential parameters on a routine period [17].

Environment and Pollution

Humans, plants, animals, and microorganisms live or work in an environment composed of land, water, and the earth's atmosphere. Four different spheres define the earth such as the biosphere (living things), the atmosphere (which includes air), the lithosphere (which includes land), and the hydrosphere (which includes water) (Fig. **2**). The chemicals and pollutants are present in a high ratio in the environment due to the fast growth of industrialization, which has become the earth's biggest problem increasing the demand for exploitation of natural resources on the earth, and causing environmental pollution in the world [18]. Some organic pollutants, organometallic compounds, inorganic ions, gaseous elements, radioactive isotopes, and nanoparticles have seriously polluted the environment. With the high density of heavy metals such as titanium, vanadium, manganese, nickel, copper, arsenic, cobalt, chromium, and molybdenum, zinc is toxic to humans, animals, and the environment.

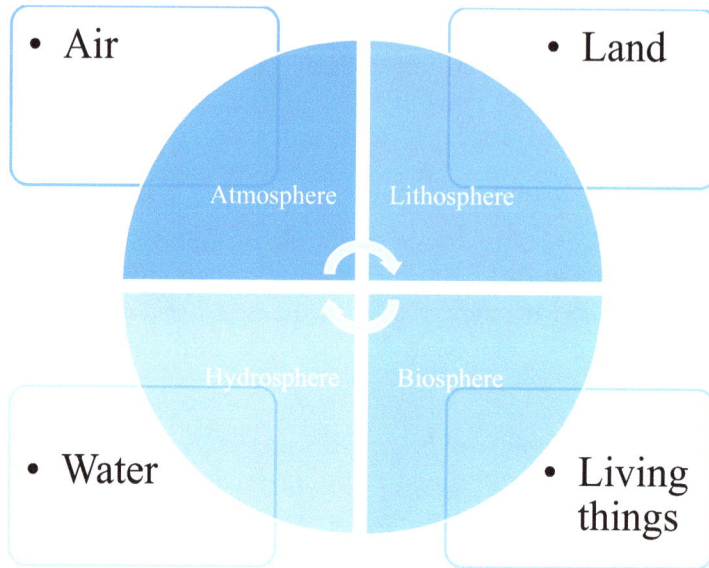

Fig. (2). Relationship of all the spheres.

Heavy metals have caused site-specific damage by interacting nuclear protein with DNA. Direct or indirect are the two types of damages that are caused by heavy metals in which conformational changes occur in biomolecules in the case of the direct method due to the heavy metals. In the case of the indirect method, because of heavy metals production by reactive oxygen and nitrogen species, they compromise with hydroxyl and another endogenous oxidant [19].

As earlier noted, microbial biosensors have significant applications in environmental monitoring; for example, in assaying the BOD, microbial biosensors have been developed to measure and detect the total content of value-related organic materials in the wastewater (Table **1**). The high consumption rate of organic matter in the wastewater by the microorganisms would measure the oxygen depletion rate interfaced with the electrodes, giving the advantage to the BOD sensors [19].

Table 1. Microbial biosensors and transducers for detecting inorganic and organic toxicities in the environment.

Target Pollutants	Microorganism	Type of Transducer	Limit of Detection	Linearity Rate Range
Mercury	*E. coli* *Chlorella*	Bioluminescence Amperometric	- -	- M to M
Zinc	*Pseudomonas putida X4*	Fluorescence	5*M	5*M to 5.5* 1 M

(Table 1) cont.....

Zinc and Copper	*E. coliXL-1 blue Circinella sp.*	Fluorescence Voltammetric	16 µM 5.4* 1 M	- 5.0* 1 to 1.0*1M
Lead	*Rhizopusarrhizus*	Voltammetric	5.0*1 M	1.0*1 M to 1.25* 15
Nitro phenols	*Pseudomonas fluorescence*	Chronoamperometric	-	-

Biosensors and their Types Used in Different Environments

The biosensor is a device used to detect signals transmitting information regarding the environment's physical and biochemical changes. To sense, the choice of biological material depends upon various factors like specificity, storage, operational, and environmental stability. However, it has been ruled out that the kind of analyte detected, such as antigens, microbes, chemical compounds, nucleic acid, and hormones, is also essential for its efficient working [20].

Microbial Biosensor Types Used in a Different Environment

Microbial biosensors are classified into three main categories based on their different signal transducers, which include electrochemical, optical, and microbial fuel cells.

Microbial Biosensors Based on Electrochemical Approach

Electrochemical microbial biosensors could provide the analyte information like specific quantitative and semi-quantitative data retained directly through spatial contact with the different electrochemical transduction elements by using biological or microbial cell components as recognition elements [21]. The different types of electrochemical microbial biosensors are summarized in Table 2.

Table 2. Different types of electrochemical transducers conjugated with microbial biosensors for specific analytes detection.

Measurement	Types of Transducer	Analyte Measured
Amperometric	Oxygen electrode	Ethanol, sugar, sucrose, surfactants, cyanide, phenolic compounds, organic matter, and organic salts.
Potentiometric	pH electrode light addressable potentiometer sensor (LAPS) oxygen electrode chloride ion-selective electrode	Drugs, urea, organic salts, amino acids, ethanol, sucrose, and trichloroethylene
Conductometric	Hybrid sensing system conjugate with transducer	Various toxic

An amperometric microbial biosensor is a type of electrochemical microbial biosensor which operates based on a reference electrode due to oxidation and reduction potentials of an electroactive species at the electrode surface. The corresponding current obtained is converted to a signal. This is because oxygen obtained is typically consumed to catalyze the enzyme reactions; the oxygen electrodes are primarily utilized. BOD biosensors are biofilm-type or whole cell-based microbial biosensors that have previously reported measuring the respiration rate of bacteria to a suitable transducer [22]. Microbial biosensors have ion-selective electrodes consisting of ammonia, pH, chloride, and other gas-sensing electrodes. These electrodes contain partial carbon dioxide (pCO_2) and partial ammonia (pNH_3) and are layered along with a microbe layer, which is immobilized in nature and is called a conventional potentiometric microbial sensor. There are several advantages of amperometric microbial sensors over potentiometric biosensors, including low relative error, high sensitivity, and a linear relationship of analyte detection with the analyte concentration and the exporting signal [23]. Thus, microbial biosensors can efficiently utilize amperometric measurements. Measurement of the conductivity changes in the medium due to the target analyte is performed by conductometric biosensors. They are sensitive in nature and provide non-specific detection of the conductance solution, which is a problem that can be bridle by detecting the differences in planer microelectronic conductance cells [24]. Another type of biosensor, microorganisms-based impedance biosensors, finds promising applications for detecting and quantifying milk and dairy products.

Microbial Fuel Cell Type Biosensors

Microbial fuel cells are bioelectrochemical transducers that have been recently focused on in research that generate electricity through an original signal using bacteria and help biodegrade the organic matter and waste products [25]. The conversation of chemical energy into electrical energy due to the microbe catalytic reaction is performed by microbial fuel cells.

The microbial fuel cell is typically a two-chamber cell with cathodic and anodic chambers separated by the proton exchange membrane (PEM). The generation of electrodes through fuel oxidation by microorganisms in the anode compartment, the electrons are transferred to the cathode compartment through the external circuit and protons are transferred to the cathode compartment through the membrane. In the cathode compartment, these electrons and protons are consumed through water formation when the proton combines with oxygen. In general, microbial fuel cells are of two types mediator and non-mediator microbial fuel cells [26]. Many microbial species have been reported and suggested to release the electrons directly through their electroactive metabolites or the anode electrode because the mediator's availability is costly and toxic, or sometimes, they are unnecessary for the process.

Based on their comparative study or analysis, mediator less or non-mediator microbial fuel cells are much studied compared to the mediator microbial fuel cell. Due to the removal of the mediator, cost from the microbial fuel cell type biosensor development technology has become more advantageous in power generation and wastewater treatment [27]. This technology is beneficial in *in situ* process monitoring, control, and pollutant analysis. BOD value monitoring primarily uses MFC-type technology as there is a proportional correlation between the BOD value and the coulombic yield of MFCs. An MFCs type BOD sensor is a microbe-enriched biosensor that can be operated for almost five years without any maintenance, if required, which can be way more service span than the other BOD sensors. However, still, some limitations include substrate concentration for BOD biosensors and more extended response time of BOD sensors. To improve the performance of MFC-type microbial biosensors, a few alternative methods have been tested for environmental monitoring. These alternatives include biocathode in two-chamber microbial fuel cells and single-chamber microbial fuel cells (SCMFC) with an air cathode [28].

Optical Microbial Biosensors

Optical principles for biosensing through transduction of biochemical interaction into a suitable output signal are called optical biosensors. The various optical properties such as adsorption, fluorescence, refractive index, and luminescence can be helpful in detecting and biosensing events [29]. Using bacterial cells as a biosensing element by optical biosensors offers advantages such as resistance to electrical noise and flexibility. Optical biosensors use optical fibres based on microbial biosensors as they are used for online monitoring, and they can be used as optical waveguides, which have been used in a large amount due to their size, flexibility, and small size [30].

Table 3. Different types of optical measurements used in microbial biosensors and their corresponding analytes.

Type of Measurement	Optical Source	Analytes Measured
Bioluminescence	Genetically modified or natural bioluminescent bacteria	Copper, arsenic, phosphorous, naphthalene, UV pollutants and toxicity, tributyltin.
Fluorescence	O_2 sensitive fluorescent material, a green fluorescent protein material	Bioavailable lysine, irons and toxins, chloroform, and BOD
Colorimetric	Photosynthetic bacteria, chromatophores	Toxin

Bioluminescence microbial biosensors are a very reliable and efficient tool because of microbial communities' light emission application which is mainly affected by the bioavailability of a fraction of detected contaminants (Table **3**). It is a very efficient environmental sensor mainly monitoring heavy metals and toxicity [31]. The genetically modified or natural bioluminescent bacteria immobilized bioluminescent system are different forms of bioluminescent bacteria used in various studies. These bacteria have shown diverse advantages such as single-chip, inexpensive component, rugged, and can be used in different types of non-laboratory [32].

The principle behind fluorescent bacteria use is that whenever an external light source is applied to the bacteria, the emission of fluorescent intensity is directly proportional to the concentration of the analyte, even at a low analyte concentration [5]. For the construction of fluorescent microbial biosensors that are sensitive to their corresponding target analytes, various fluorescent components like green fluorescent protein and O2 fluorescent material have been used widely. Bioluminescence can detect the sensitive target analyte [33].

Colourimetric measurements have not been widely focused on because they depend on colour changes, and few literature surveys have been found on compared studies until now. For example, a cytosensor, a sensitive biosensor with high sensitivity to toxins, is based on chromatophores recently developed to determine or monitor the fish cell and microbes' interaction in toxin-sensitive living cells [34]. The photosynthetic bacteria is used as a sensor strain to construct a novel type of colourimetric whole-cell biosensor in which it is notified that the change in bacteria colour from green to yellow is more explanatory than the carotenoid-based whole-cell biosensor reporter signal [35].

Monitoring Organic Compounds

Urbanization and industrialization have put immense pressure on the environment. Many industrial and sewage waste is added to the river streams without proper treatment. There is an increment in eutrophication due to increased concentration of phosphorous, Nitrogen, algal blooms, and many other organic compounds in the water. It has led to the depletion of water resources, affecting aquatic life (Fig. **3**). Several studies have shown that agriculture pollution also increases the nutrients loading, thus, contributing to increasing BOD. Irrigation, fertilizers, drainage, and crop residues also flow into the river along with rainwater. Nutrient enrichment is associated with phosphorus and Nitrogen from agriculture that can disturb the oxygen level and affect the ecosystem. This also disturbs the aquatic flora and fauna. Managing water quality is one of the important factors in controlling water pollution. Many standards have been established to check water quality, such as chemical oxygen demand, dissolved oxygen, and biological oxygen demand [36]. Dissolved oxygen depends on various factors, such as temperature, salinity, and oxygen depletion. COD is the oxygen required for the chemical oxidation of organic matter in the presence of chemical oxidants [37]. The BOD defines the quality of the water environment and organic water pollution. Biological oxygen demand is the amount of oxygen microorganisms need to break down the organic matter in the water sample.

BOD knew to have three major applications:

1. It helps in assessing the extent of water pollution.
2. The ratio between BOD and COD gives us information about the required size of the wastewater treatment plant.
3. The BOD and COD ratio indicates the biodegradable effluent fraction [38].

Table 4. Assessment of biological oxygen demand of some rivers in 2010 and 2011 [39].

River	BOD mg/l (2011)	BOD mg/l (2010)	Trend
Godavari	37.0	60.0	Decreased
Gomati	10.5	12.2	Decreased
Bindusar	7.4	7.0	Increased
Mahananda	6.6	5.5	Increased
Pennar	6.0	4.4	Increased
Satluj	32.0	40.0	Decreased
Brahmaputra	9.2	6.3	Increased
Chambal	42.0	48.0	Decreased
Tapi	10.0	16.0	Decreased
Dwarka	12.2	15.4	Decreased
Ganga	11.0	15.0	Decreased
Krishna	16.0	10.0	Increased
Sirsa	15.0	8.0	Increased
Churni	64.0	3.7	Increased
Savitri	525.0	5.4	Increased
Ram Rekha	15.0	3.5	Increased
Pedhi	46.0	16.4	Increased

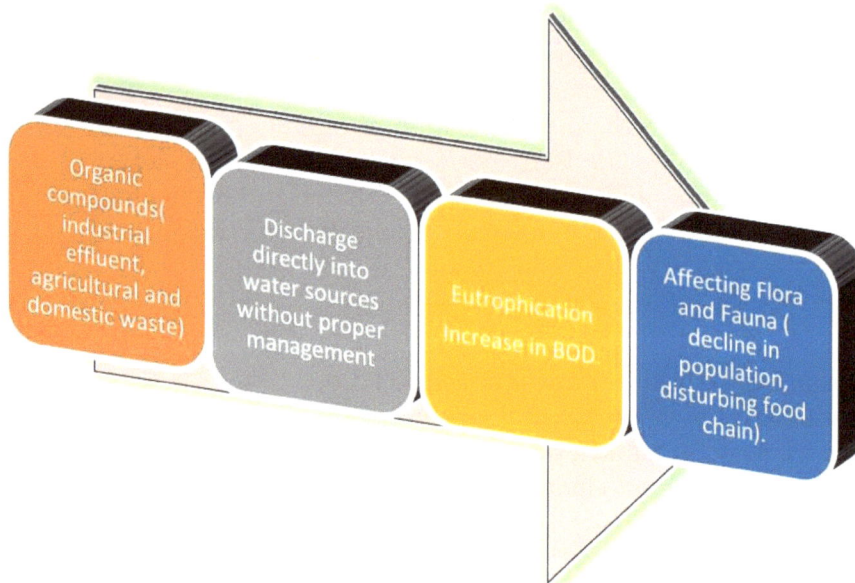

Fig. (3). Shows the link between organic compounds and aquatic life.

There is a significant increase in BOD in Pedi, Savitri, and Churni (Table **4**). The increase in BOD observed is due to the increase in the discharge of sewage and other effluents into the river without proper treatment and management.

As per NWMP (National water Quality Monitoring program), polluted river stretches have been divided into five priority classes depending on their BOD concentration and exceeding concentration identified as polluted (Table **5**).

Table 5. NWMP (National water Quality Monitoring program) based on five classes depending on BOD concentration.

Categories	BOD Concentration (mg/L)
Type-1	More than 30
Type-2	20-30
Type-3	10-20
Type-4	10-6
Type-5	6-10

BOD exceeding three mg/l will not meet the water quality requirement, but it will not affect the dissolved oxygen in the water level. If BOD exceeds six mg/l, the dissolved oxygen level will drop as shown in Table **6** [40].

Table 6. State-wise number of polluted rivers identified in 2018 [40].

State / Union Territory	Polluted River Stretched
Andhra Pradesh	5
Assam	44
Bihar	6
Chhattisgarh	5
DD and DNH	1
Delhi	1
Goa	11
Gujarat	22
Haryana	2
Himachal Pradesh	7
Jammu & Kashmir	9
Jharkhand	7
Karnataka	17
Kerala	21
Madhya Pradesh	22
Maharashtra	53
Manipur	9
Mizoram	9
Meghalaya	7
Nagaland	6
Odisha	19
Punjab	4
Rajasthan	2
Puducherry	2
Sikkim	4
Tamil Nadu	6
Telangana	8
Tripura	6
Uttarakhand	9
U. P	12

As per the report of the oversight committee in compliance with the National Green Tribunal about Water quality and drain status in Uttar Pradesh, the place from where Ganga enters district Bijnor the BOD was found to be 1.9 mg/l in 2019 which means water is fit for bathing. However on entering Kannauj, the BOD changes to 3.3mg/l. Similarly, the BOD of the Yamuna in Noida was found to be 50 .8mg/l. Similarly, at Vrindavan and Mathura, water quality was D. This is the malfunctioning of the drainage system and discharge of industrial effluents directly into the river without proper treatment. According to a survey done in India and Africa, the amount of nitrate found in wells is greater than 50mg/l and, in some cases, 100mg/l as per Pacific institute, World Water quality Facts, and Statistics. The water quality of some Indian rivers are mentioned in Table 7.

Table 7. Water quality of some river (BOD, Faecal Coliform, nitrate) in the year 2017) [41].

River	BOD <3.0mg/L		Faecal Coliform Range (<2500 MPN/100ml)		Total Coliform Formed (MPN/100ml max)	Nitrate	
	Minimum	Maximum	Minimum	Maximum		Minimum	Maximum
Sutlej at 100m d/s Buddha nala confl., Ludhiana	30	108	17000	210000	380000	1.6	1.4
Beas at g.t. road Under bdg. Near Kapurthala	1	1.6	70	140	280	0.8	1.9
Sirsa river, d/s Nalagarh bridge	4	16	14	38	210	0.2	4.8
Banganga (bathing Ghat), katra,	2	14	-	-	-	-	-
Tawi below tawi Bridge	2.1	8.3	-	-	-	0.2	7.3
Ganga at Haridwar D/s	1	6.6	-	-	1600	-	-
Yamuna at okhla After meeting of Shahdara drain, Delhi	8	80	2600	22000000	22000000	0	0
Yamuna at khojkipur Panipat	1	55	33	4700	160000	0.01	1.01
Bhella river at Lohiya bridge d/s Kashipur	12	64	-	-	-	-	-
Gomti at Lucknow D/s, U.P.	7	14	54000	170000	210000	0.3	3

(Table 7) cont.....

Brahmaputra river at Dhenukhapahar, Assam	0.6	6.4	360	2100	110000	0.3	2.4

The standard procedure to measure the BOD is a five-day method. It is quite complex and time-consuming. The complexity of this method led to the development of a BOD biosensor. Biofilm biosensor with oxygen electrodes measures the respiration rate of microorganisms in the transducers. Such biosensors consist of microbial film to be used as a biorecognition element. Apart from this, other BOD biosensors are available, such as mediator biosensors, optical biosensors, microbial biofuel cells as BOD biosensors, *etc.* Different strains of microorganisms, such as *B. subtilis, Pseudomonas putida, etc.,* have been used as a biorecognition element in BOD microbial biosensor [42].

Studies have shown that Artificial Neural Network models demonstrate the dissolved oxygen value in river water. Water temperature and five days-biological oxygen demand were used to prepare the feed-forward neural network (FNN) using the backpropagation studying algorithm system that has shown a significant correlation coefficient (0.885) and efficiency coefficient (0.782), suggesting that water controller could employ these results to manage the water treatment plant [43]. Research conducted in 2016 has shown that open-type biosensors can be used for *in situ* monitoring of BOD in anaerobic conditions. This biosensor inserted the open-type anode into a tank containing livestock wastewater. Potentiostat regulated the anodic potential. Biosensor showed a BOD of up to 250 mg/L and a logarithmic correlation (R2 > 0.9) [44]. Studies have shown that using an electrical coagulation process, iron, and aluminum rods can remove BOD and COD from greywater. DR/5000UV–vis HACH spectrophotometer was used to determine the BOD and COD. Various chemicals, such as H_2SO_4, $K_2Cr_2O_7$, $HgSO_4$, and Ag_2SO_4, were added to water and stored in the dark at 4 °C for this metal analysis. This cost-effective method can be widely used in wastewater treatment to remove pollutants [45].

The use of microbial fuel cells for wastewater treatment, environmental monitoring, and electricity production is on-trend. Single-chamber Microbial Fuel Cells (SMFCs) are widely used as biosensors made by using Andean soil, which monitors BOD in which each one is monitoring a BOD concentration of 10, 100, and 200 mg/L of synthetic washed rice wastewater named SMFC1, SMFC2, and SMFC3. The result obtained consisted of three stages transient, growth, and stable in the SMFCs. SMFC2 needed more time to reach a steady stage than SMFC1, thus showing that BOD influenced the factors in SMCFS. The OCV ratios were found to lie between 40.6–58.8 mV and 18.2–32.9 mV in the case of SMFC1 and

SMFC2 [46]. Feed-forward and radial basis function neural networks based on artificial intelligence were used to determine the BOD in the sefidrood river. The results show that FFANN and RBFANN models work efficiently and show BOD5(max) R values near 0.89 and 0.90. It was also found that FFANN and RBFANN show almost the same performance for the detection.

A BOD biosensor was used to monitor the organic compound in domestic wastewater. 282 ± 23 mA/m2 current density was found in domestic wastewater. A linear relationship was obtained between current density and BOD concentration ranging from 17 ± 0.5 mg O2/L to 78 ± 7.6 mg O2/L. These results give us helpful information about developing biosensors that can be used for *in situ* monitoring of wastewater [47]. The study was conducted to predict the amount of Biochemical Oxygen Demand (BOD) in the Chaophraya River of Thailand using the Alpha-Trimmed ARIMA Model and the data collected in the past. The obtained results were then compared with the other three existing models and the results obtained have fewer relative errors than half of the other three models' other relative errors. The accuracy shown by this model was more than 70% [48].

In 2019, a study predicted the DO, BOD, and COD of Selangor River of peninsular Malaysia. The root of this model was vector autoregression (VAR). Adopted stimulation was based on the three inputs:

   i. Parameters of Water Quality.
  ii. Water quality parameters and river flow data.
 iii. Water quality parameters and rainfall data.

It was found that with the increase in input number, the prediction accuracy of the VAR model substantially increased. This model predicted the accuracy where all the variables such as hydrological, environmental, and climatological were considered. The results have shown that the VAR model can provide better information to the higher authorities to take steps against the quality of river water [49].

A biosensor was made to determine the BOD of industry effluent by immobilizing the microbial consortium onto the cellulose acetate membrane in closeness to the DO electrode. The consortium was collected from effluent, and glutamic/ glucose solution was used as a standard for calibrating a biosensor. The response time of the biosensor was optimized and used to determine the BOD of the effluent. The response time showed a detection limit of 1mg/ml for around seven minutes. Virtuous linear range *RR2* 0.99 with relative standard deviation (RSD) < 9% was observed with a glutamic/glucose acid standard solution [50]. Bioluminescent bioreporter pad biosensor is portable biosensors consisting of disposable and non-disposable parts. The disposable part consists of a calcium alginate matrix with immobilized bacteria, and the non-disposable part consists of (a CMOS photodetector). A variety of toxicants were tested against biosensors. The results have shown that CMOS biosensors possess higher sensitivity. It is also user-friendly and has a lower detection limit, and the amount of sample required is very low, thus making it suitable for testing the water quality of underdeveloped areas [51].

Monitoring Pathogen

A pathogen is a microorganism that causes or can cause disease. These can be opportunistic or non- opportunistic [52]. Pathogens adapt themselves according to their changing environment by changing their base sequence and thus gene expression. The presence of fungi, bacteria, and parasitic pathogens in the environment threatens public health (Table **8**). Therefore, specific methods are required to detect and monitor pathogens in food, air, water, and soil.

A biosensor must be specific and sensitive to detect the even low concentration of pathogens in treated and untreated water.

Biosensors that detect pathogens mainly consist of capture molecules, a labeled antibody is interacting with bacteria, and a signal detector. The recognition element recognizes the element, and sensitivity is obtained through signal transduction (Table **9**).

Table 8. Pathogens and their associated disease [53, 54].

Pathogen	Disease
Salmonella typhi	Typhoid fever
E. coli	Gastroenteritis, Hemolyticuremia
Adenoviruses	Gastroenteritis, respiratory Infection
Rotavirus	Gastroenteritis
E. histolytica	Amoebic dysentery
P. aeruginosa	Pulmonary disease, skin infection
Vibrio cholera	Cholera
Flavivirus	Dengue
Trypanosoma brucei	African trypanosomiasis
Rickettsia typhi	Murine typhus
Campylobacter spp.	Campylobacteriosis
Acanthamoeba	Corneal ulcers

Table 9. Different biosensors used for detection of different analytes.

Type of Biosensor/Measurement	Analyte Detected	References
Fluorescence	Iron, toluene, lysine, arsenate, galactosides, chloroform	[55]
Amperometric	BOD, sucrose, p-nitrophenol, surfactants, cyanide	[55]
Oxygen electrode	BOD	[56]
Carbon nanotube	*Staphylococcus aureus*	[57]
Gold nanoparticle	*Phytophthora infestans*	[58]
Optic fiber	*Salmonella* enterica	[59]
Liquid crystal based	*Erwinia carotovora, Rhazictonia Solani*	[60]
	Ganoderma boninense	[61]
Waveguide-based	*Mycobacterium*	[62]
Piezoelectric biosensor	*Escherichia coli* O157.H7	[63]
***Pectobacterium atrosepticum* biosensor**	Blackleg and Soft Rot Disease of Potato	[64]
Yeast biosensor	*C. albicans* and *H. capsulatum*	[65]
Gold nanoparticle	*Escherichia coli* O157.H7 and *Salmonella typhimurium*	[66]
Microelectrode based impedance biosensor	*Salmonella enteritidis*	[67]
Electrochemical biosensor	*Pathogenic viruses*	[68]
Electrochemical and optical biosensors	*Campylobacter* and *Listeri*	[69]
Magnetoelastic biosensors	*Salmonella typhimurium*	[70]

Nanoparticle-based biosensors are very interesting these days due to their specificity and sensitive detection of biotic and abiotic contaminants.

Such types of biosensors must have the following characteristics:

1. Should be able to distinguish between pathogenic and non – pathogenic organisms.
2. Should be able to detect even a small quantity of target, thus promoting non-specific binding (binding between antibody and non-antigenic material) [71].

These sensors use nanoparticles as signals to increase the output. *Cryptosporidium parvum* is a water-borne pathogen that contaminates the drinking water, leading to many diseases. The assay detected this pathogen at a limit of 10-1000 organisms per cell [72].

Impedimetric biosensors have gained much importance these days in bacteria detection. It is an electrical biosensor that can detect foodborne pathogens efficiently as it is sensitive and performs on the spot [73].

For detecting *Listeria monocytogenes*, a foodborne pathogen, a PCR assay was applied to the spike and natural paneer samples. The assay's sensitivity was 10^4 cells before enrichment on the L. monocytogenes spiked paneer. After four hours of enrichment, the improvement observed was 10^3 cells, and after 6 h of enrichment, one to 10. When applied to ten random samples, no amplified product was observed, thus showing the absence of *L. monocytogenes* [74].

Paper-based Analytical Microfluidic Devices(µPADs) are also preferred over conventional methods for detecting pathogens. For the detection of lipopolysaccharides of *Salmonella typhimurium* and *Salmonella enteritidis,* chemiluminescent immunoassay 59 was used. A paper-based lateral flow strip was used for analysis which is made up of four different membranes that arrest the pathogen and detect it. Liquid samples are pumped through paper substrates. This method is better than the conventional one because of its low cost, ease of use, and flexibility. These are thin, disposable, and easy to fabricate.

Surface plasma resonance biosensors recognize microorganisms in liquid or food dilutions. SPR measures the change in the refractive index. The evanescent wave interacts with the metal film's free electrons at the Surface plasma resonance angle. SPR reflectance curve can be recorded at different angles of incidence of light. A study has shown that using SPR immunosensor miniaturized with a microfluidic system, a 10,000 CFU/mL cell concentration of *L. pneumophila* was identified [75].

A graphene-based biosensor system has a high sensitivity to virus pathogen detection. The electrochemical sensor was prepared in two steps:

1. Synthesizing graphene oxide from GO colloidal suspensions using vacuum concentrator and thermal annealing method to produce reduced G2 film.
2. Modifying the graphene surface with the pyrene derivative and then linking it covalently with virus-specific antibodies.

Graphene film arrests the Rotavirus through antigen-antibody interaction with high sensitivity and selectivity [76]. Surface Plasma resonance using lectin as a receptor is used to monitor the *E. coli* O157. H7 in food samples. Lectins discovered from *T.vulgaris, C. ensiformis, U. europaeus, A.hypogaea,* and *M. amurensis,* were used to estimate the selectivity of the approach for binding *E. coliO157. H7* effectively. Results have shown that these biosensors were sensitive, consistent, and effective for recognizing *E. coli O157. H7* and can be employed in the food industry [77]. Understanding the immune system's response to pathogen environment attacks and monitoring such processes need well-defined techniques. A microfluidic device coupled with a worm-based biosensor can be used to study such a response. Worms stored in the chambers were exposed to pathogens. During the pathogen infection time- a lapse was measured. Results have shown that in the presence of *P. aeruginosa* strain PA14, an irg-1 gene could be induced as well, as, under antibiotics, gentamicin and erythromycin infected worm survival could be rescued. Thus, providing a good field to study pathogenesis research [78]. *E. coli* affects the intestine by producing heat-stable and reliable toxins. The Quantum dot technique has been reported to be applicable against *E. coli* detection as its surface has mannose-specific lectin. CDS quantum dots get coated with thiolate mannose that aggregates with *E. coli.* The absence of aggregation with strains that have capping other than thiolate mannose shows that the assay is very specific [79].

Monitoring Inorganic Compounds

The heavy metals and metalloids like arsenic, cadmium, chromium, nickel, mercury, and lead constitute a significant part of inorganic compounds disposed of in the environment, and their remediation is globally important. The Agency for Toxic Substances and Disease Registry (ATSDR) was formed by the Comprehensive Environmental Response, Compensation, and Liability Act (CERCLA) to prioritize substances that were determined to be the most significant threat to human health and the environment (Table **10**).

Table 10. The rank of inorganic pollutants which are listed in the Agency for Toxic Substances and Disease Registry (ATSDR) 2019.

Rank (as per 2019)	Substance Name	Safety Level Limit in Water	Safety Level Limit in Air	Safety Limits in Soil
1	Arsenic	0.01 ppm	10 $\mu g/m^3$ as arsenic dust	20 ppm
2	Lead	0.05 ppm	0.15 $\mu g/m^3$	400 ppm
3	Mercury	0.002 ppm	0.05 mg/m^3 as mercuric vapor	722 ppm
7	Cadmium	0.005 ppm	200 $\mu g/m^3$	0.003 ppm
35	Cyanide	0.2 ppm	5 mg/m^3 as hydrogen cyanide	0.1 ppm

The presence of such inorganic pollutants above the safety level in the environment is due to various anthropogenic activities, and high concentration in the environment leads to bioaccumulation in the food chain. Humans at the highest feeding level are more prone to get health compromises. The different organs affected by heavy metal poisoning are summarized in Fig. (4). Industrial discharge, chemical-based fertilizers, and pesticides are sources of inorganic pollutants that contaminate the environment. So, it becomes crucial to monitor the inorganic pollutants, especially heavy metals, to maintain a safe level for the benefit of the environment. Such heavy metals or inorganic compounds often cause short and long-term effects on plants, animals, and human communities that are evident in the terrestrial environment. Also, the marine or aquatic ecosystem is equally at high risk, which could concentrate such inorganic compounds in the range of milligram per liter through biological magnification. The varied effects like deficient growth, photosynthesis, low reproduction rate, behavioral changes, or high death rate could be possible in many cases depending on the duration and the exposure.

Earlier, conventional analytical techniques, such as atomic or mass or emission spectroscopy and gas chromatography, were used to detect or sense environmental pollutants. However, these methods were not feasible in small laboratories or developing countries due to expensive instruments, tedious sample preparation, time-consuming work, and transportation cost. The usefulness of the techniques was also limited as they provided less information depending on the bioavailability of the heavy metal ions. An alternative to these methods, biosensors have various applications like *in situ* analyte sensing and rapid result [80]. For detecting inorganic heavy metals, enzyme-based or whole-cell biosensors are used. In enzyme-based biosensors, enzymes that correspond to the analyte used are immobilized in the device, whereas whole-cell uses microbial cells that sense any environmental fluctuation like pH, temperature changes, nutrient availability, and toxic compounds. Synthetic biologists exploit this function of microbes in designing biosensors with higher sensitivity and specificity. Many whole-cell biosensors are utilized for detecting or sensing various environmental heavy metal pollutants. The microbial cells have reporter genes that might be native or engineered using various genetic engineering tools that behave as bioreceptor for the analyte. Once the analyte has been exposed to the bioreceptor, fluorescence or luminescence biochemical reactions occur depending upon the nature of the bioreceptor. Such a signal is passed to a transducer connected to the microbial cell on one end and a reading device on another end [81].

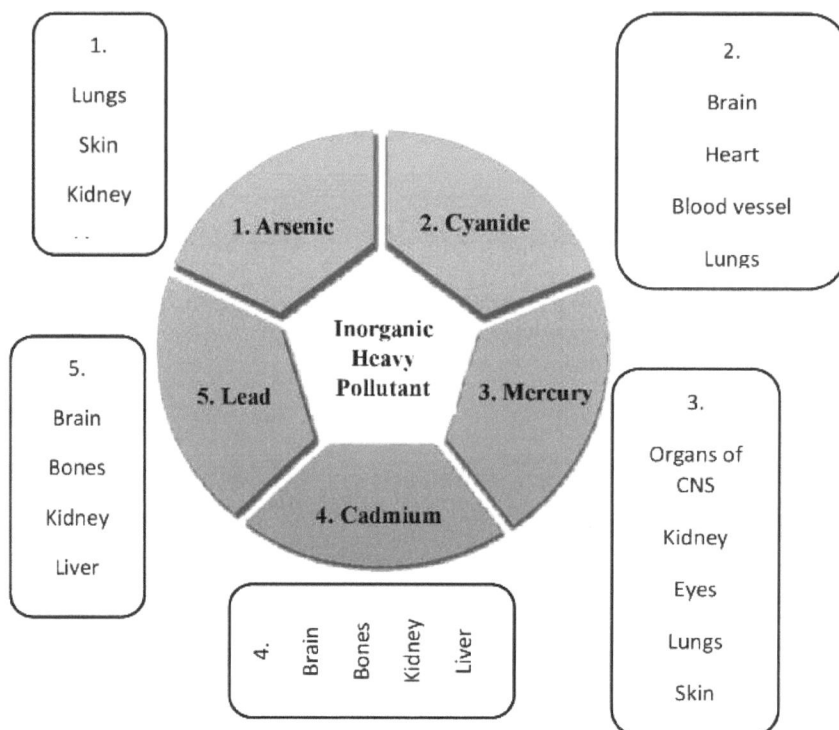

Fig. (4). Inorganic heavy pollutants affecting major organs of a human body.

The common mercury pollutants found in the environment are elemental mercury (Hg), mercuric chloride ($HgCl_2$), mercurous chloride (Hg_2Cl_2), mercuric ammonium chloride ($Hg(NH)_2Cl_2$) and methyl mercury (CH_3Hg) [82]. The presence of mercury harms the wildlife ecosystem. The terrestrial and aquatic food chains are also affected, compromising the health of top consumers like humans. The coordination misbalance of the nervous system, compromised immune system, molecular gene alteration, gastrointestinal damage, and kidney failure are some of the problems humans encounter. It is more important to monitor the bioavailability of metals than to estimate total metal concentration to assess the risk factor. The whole-cell biosensors are used for the detection of mercuric pollutants. The main components required for constructing biosensors are *mer* operon having regulatory elements fused with *lux* gene isolated from *Vibrio fischeri*. These fused *mer-lux* gene elements are inserted into *Escherichia coli*, which has heat shock promoters. This type of biosensor measures mercuric pollutants through bioluminescence which is detected and quantified using a luminometer as a relative luminescence unit (RLU) [83]. The figure explains that the underlying mechanism for using mer-lux biosensors is that the *E. coli* cell membrane passively allows mercuric ions (Hg^{2+}) to enter the cell.

Once inside, the cytoplasm, Hg^{2+} ions enter the cell's nucleus, which binds to the promoter mer promoter and subsequently allows higher transcription of merR gene elements and lux genes. The lux transcripts further encode luciferase enzyme, which binds to the substrate, and the product bioluminescence, which in turn passes through a silicon-based photomultiplier photodetector that collects the incoming bioluminescence signals. The most common toxic arsenic forms found in the environment are As^{3+} and $As^{5+,}$ and ArsR-based biosensors are available to detect arsenic. ArsR is a responsive transcriptional repressor molecule bound to the operator/promoter site of the operon in the absence of arsenite, which does not allow biosynthesis of *ars* genes. So, the biosensors are constructed using the *ars*R gene coupled with reporter genes like *lac*Z, *lux*, and *gfp*. In the presence of arsenite, the reporter genes are expressed, producing specific enzymes and lowering the pH. These negative auto-regulated consequences have been detected as fluorescence, luminescence, or absorbance. The expression of reporter genes is analyzed as luminescence, fluorescence, or absorbance [84 - 86].

The following heavy metal that needs to be sensed because of the various adverse environmental impacts is cadmium. The biosensors designed for sensing cadmium depend on the coupling of the sensing elements' cadmium (Cd) activated promoter with target or reporter genes in genetically altered microorganisms [87, 88]. The different range of whole-cell biosensors used for cadmium detection has been reported and listed in the table. The lead sensing biosensors have been constructed through the pbrR promoter, which will bind to the lead (Pb) and lead to the expression of the reporter gene coupled to the promoter genes [89]. Lastly, cyanide also needs to detect through whole-cell biosensors due to its extremely harmful effects on human health. However, unlike other heavy metals detected using specific responsive elements, cyanide detection occurs in an unusual approach. Constructing the microbial or whole-cell biosensors depends upon measuring the respiratory activity of the cell. The principle behind this measurement is that in the presence of cyanide, the activity of cytochrome oxidase is inhibited, which decreases the efficiency of the respiration process. Most biosensors are connected to an oxygen electrode that detects the current generated due to oxygen concentration. Such biosensors have immobilized microorganisms capable of performing oxidative respiration, which is important for monitoring cyanide pollutants in the water (Table **11**). In another approach, many microorganisms can produce cyanidase (cyanide hydrolase) enzyme, which cleaves cyanide into formic acid and ammonia. Such microorganisms are immobilized in the construction of biosensors where the sensors are designed to detect formate and convert it into carbon dioxide (CO_2) utilizing Nicotinamide adenine dinucleotide (NAD^+). Hence, NAD^+ produced will be converted to β-Nicotinamide adenine dinucleotide hydrate ($NADH^+$) using oxygen. This oxygen consumption is measured by the current produced through the oxygen electrode [90 - 93].

Table 11. Summary of whole-cell biosensors used for sensing inorganic pollutants.

Sr. No.	Heavy Metal Detected	Promoter/Reporter Construct	Detectable Output Signal	Detectable Range	Host	References
1	Mercury	merR/PpFbFp genes	Fluorescence	1-2 nM	*Escherichia coli* NEB5α	[94, 95]
2	Mercury	merR/lux genes	luminescence	0.25 µg	*Escherichia coli* DH5α	[96]
3	Mercury	merR/lux genes	Luminescence	0.05-0.5 nM	*Escherichia coli* JM109	[83]
4	Mercury	merR/gfp genes	Fluorescence	1 nM	*Escherichia coli* BL21 (DE3)	[97]
5	Mercury	merR/rfp genes	Fluorescence	1 nM	*Escherichia coli* DH5α	[98]
6	Mercury	merR/lux genes	luminescence		*Escherichia coli* CM2624	[99]
7	Arsenic	arsR/lux genes	Fluorescence	0.1 µM	*Escherichia coli* DH5α	[100]
8	Arsenic	arsR/gfp genes	Fluorescence	0.1-4 µM	*Escherichia coli* DH5α	[85]
9	Arsenic	nikR/egfp genes	Fluorescence		*Escherichia coli* DH5α	[44]
10	Arsenic	arsRBS2/mcherry genes	Fluorescence	25-800 nM	*Escherichia coli* NEB10	[101]
11	Arsenic	arsR/gfpuv	Fluorescence	10 µg/L	*Escherichia coli* MG1655	[102]
12	Arsenic	arsR/lux genes / arsR/venus genes	Luminescence / Fluorescence	10 nM / 10 nM	*Magnetospirillium magneticum* AMB-1 *Magnetospirillium gryphiswaldense* MSR-1	[103]
13	Cadmium	zntR/gfp	Fluorescence		*Escherchia coli* K-12 MG1655	[104]

(Table 11) cont.....

14	Cadmium	(1) DR_0659/lacZ genes (2) DR_0659/crtI genes	Absorbance	(1) 10-1 mM (2) 50 nM–1mM	*Deinococcus radiodurans*	[87]
15	Cadmium	groEL/lacZ genes	Absorbance	3 μg/mL	*Escherichia coli*	[105]
16	Cadmium	zntR/lacZ genes	Electrochemical	5.0 μM (soil) 25 nM (water)	*Escherichia coli* RBE23-17	[106]
17	Cadmium	cadA/lux	Luminescence	10 nM 3.3 nM	*Staphylococcus aureus* RN4220 (pTOO24) *Bacillus subtilis* BR151 (pTOO24)	[107]
18	Lead	pPpbr/vio genes	Absorbance	0.1875 μM	*Escherichia coli* TOP10	[108]
19	Lead	pbrR/gfp	Fluorescence	0-6 μg/mL	*Escherichia coli* DH5α	[89]
20	Lead	pbrR/gfp	Fluorescence	50-400 μM	*Escherichia coli* DH5α	[109]
21	Lead	cadA/lux	Luminescence	33 μM	*Staphylococcus aureus* RN4220 (pTOO24)	[107]
22	Cyanide	-	Voltage	0.5-150 μM	*Saccharomyces cerevisiae*	[91]
23	Cyanide	-	Voltage	2.31 μM	*Flavobacterium indicum*	[93]
24	Cyanide	-	Current	0.5 μM	*Thiobacillus ferrooxidans*	[110]
25	Cyanide	-	Current	7.3 μMol/L	*Pseudomonas* spp.	[92]

CONCLUSION

Microorganisms are best suited for the detection or sensing of environmental pollutants. Various sensing elements of microorganisms have been exploited for specific kinds of pollutants. However, the recent development in the field of genetic engineering and recombinant techniques have widely improved the efficiency of microbial biosensors. The advantages over conventional type sensors have been discussed above. However, most of the research carried out using microorganisms for developing biosensors, and its application is still in *in-vitro* condition. The utilization of microbial biosensors needs to be carried out in field studies in the coming future. The importance of more diverse recognition elements or sensing elements of the microorganisms and the search for more efficient transducers need further exploitation. The advancement in new materials and technologies for developing biosensors needs further research for a promising bright future. Lastly, proper commercialization needs to be worked for more awareness and advantages of microbial biosensors over chemical biosensors.

CONSENT FOR PUBLICATION

Not applicable.

CONFLICT OF INTEREST

The authors declare that there is no conflict of interest.

ACKNOWLEDGEMENTS

The author acknowledges Lovely Professional University, Punjab, for providing the necessary facilities and support in formulating the manuscript.

REFERENCES

[1]   Cella LN, Sanchez P, Zhong W, Myung NV, Chen W, Mulchandani A. Nano aptasensor for protective antigen toxin of anthrax. Anal Chem 2010; 82(5): 2042-7.
      [http://dx.doi.org/10.1021/ac902791q] [PMID: 20136122]

[2]   Wu CH, Le D, Mulchandani A, Chen W. Optimization of a whole-cell cadmium sensor with a toggle gene circuit. Biotechnol Prog 2009; 25(3): 898-903.
      [http://dx.doi.org/10.1002/btpr.203] [PMID: 19507257]

[3]   Mohanty SP, Kougianos E. Biosensors: a tutorial review. IEEE Potentials 2006; 25(2): 35-40.
      [http://dx.doi.org/10.1109/MP.2006.1649009]

[4]   D'Souza SF. Immobilization and stabilization of biomaterials for biosensor applications. Appl Biochem Biotechnol 2001; 96(1-3): 225-38.
      [http://dx.doi.org/10.1385/ABAB:96:1-3:225] [PMID: 11783889]

[5]   Lei Y, Chen W, Mulchandani A. Microbial biosensors. Anal Chim Acta 2006; 568(1-2): 200-10.
      [http://dx.doi.org/10.1016/j.aca.2005.11.065] [PMID: 17761261]

[6]   Su L, Jia W, Hou C, Lei Y. Microbial biosensors: A review. Biosens Bioelectron 2011; 26(5): 1788-99.
      [http://dx.doi.org/10.1016/j.bios.2010.09.005] [PMID: 20951023]

[7]   Tuncagil S, Ozdemir C, Demirkol DO, Timur S, Toppare L. Gold nanoparticle modified conducting polymer of 4-(2,5-di(thiophen-2-yl)-1H-pyrrole-1-l) benzenamine for potential use as a biosensing material. Food Chem 2011; 127(3): 1317-22.
      [http://dx.doi.org/10.1016/j.foodchem.2011.01.089] [PMID: 25214132]

[8]   Choi S, Chae J. Optimal biofilm formation and power generation in a micro-sized microbial fuel cell (MFC). Sens Actuators A Phys 2013; 195: 206-12.
      [http://dx.doi.org/10.1016/j.sna.2012.07.015]

[9]   Du Z, Li H, Gu T. A state of the art review on microbial fuel cells: A promising technology for wastewater treatment and bioenergy. Biotechnol Adv 2007; 25(5): 464-82.
      [http://dx.doi.org/10.1016/j.biotechadv.2007.05.004] [PMID: 17582720]

[10]   Erbe JL, Adams AC, Taylor KB, Hall LM. Cyanobacteria carrying ansmt-lux transcriptional fusion as biosensors for the detection of heavy metal cations. J Ind Microbiol 1996; 17(2): 80-3.
[http://dx.doi.org/10.1007/BF01570047] [PMID: 8987894]

[11]   Reshetilov AN, Iliasov PV, Knackmuss HJ, Boronin AM. The nitrite oxidizing activity of Nitrobacter strains as a base of microbial biosensor for nitrite detection. J. Anal Lett 2000; 33(1): 29-41.
[http://dx.doi.org/10.1080/00032710008543034]

[12]   Ramsay G. Commercial Biosensors: Applications to Clinical, Bioprocess and Environmental Samples. Chichester, UK: Wiley 1998.

[13]   Riedel K. Microbial biosensors based on oxygen electrodes. Enzyme and Microbial Biosensors. Human Press 1998; pp. 199-223.
[http://dx.doi.org/10.1385/0-89603-410-0:199]

[14]   Rogers KR. Recent advances in biosensor techniques for environmental monitoring. Anal Chim Acta 2006; 568(1-2): 222-31.
[http://dx.doi.org/10.1016/j.aca.2005.12.067] [PMID: 17761264]

[15]   Steinberg SM, Poziomek EJ, Engelmann WH, Rogers KR. A review of environmental applications of bioluminescence measurements. Chemosphere 1995; 30(11): 2155-97.
[http://dx.doi.org/10.1016/0045-6535(95)00087-O]

[16]   Belkin S. Microbial whole-cell sensing systems of environmental pollutants. Curr Opin Microbiol 2003; 6(3): 206-12.
[http://dx.doi.org/10.1016/S1369-5274(03)00059-6] [PMID: 12831895]

[17]   Malhotra BD, Chaubey A. Biosensors for clinical diagnostics industry. Sens Actuators B Chem 2003; 91(1-3): 117-27.
[http://dx.doi.org/10.1016/S0925-4005(03)00075-3]

[18]   Masindi V, Muedi KL. Environmental contamination by heavy metals. J Heavy Metals 2018; 10: 115-32.

[19]   Walker CH, Sibly RM, Hopkin SP, Peakall DB. Principles of Ecotoxicology. CRC press 2012.

[20]   Zhang Y, Angelidaki I. A simple and rapid method for monitoring dissolved oxygen in water with a submersible microbial fuel cell (SBMFC). Biosens Bioelectron 2012; 38(1): 189-94.
[http://dx.doi.org/10.1016/j.bios.2012.05.032] [PMID: 22726635]

[21]   Thévenot DR, Toth K, Durst RA, Wilson GS. Electrochemical biosensors: recommended definitions and classification. Biosens Bioelectron 2001; 16(1-2): 121-31.
[PMID: 11261847]

[22]   Liu J, Mattiasson B. Microbial BOD sensors for wastewater analysis. Water Res 2002; 36(15): 3786-802.
[http://dx.doi.org/10.1016/S0043-1354(02)00101-X] [PMID: 12369525]

[23]   Bhatia R, Dilleen JW, Atkinson AL, Rawson DM. Combined physico-chemical and biological sensing in environmental monitoring. Biosens Bioelectron 2003; 18(5-6): 667-74.
[http://dx.doi.org/10.1016/S0956-5663(03)00012-5] [PMID: 12706577]

[24]   Guan JG, Miao YQ, Zhang QJ. Impedimetric biosensors. J Biosci Bioeng 2004; 97(4): 219-26.
[http://dx.doi.org/10.1016/S1389-1723(04)70195-4] [PMID: 16233619]

[25]   Davis F, Higson SPJ. Biofuel cells—Recent advances and applications. Biosens Bioelectron 2007; 22(7): 1224-35.
[http://dx.doi.org/10.1016/j.bios.2006.04.029] [PMID: 16781864]

[26]   Aelterman P, Rabaey K, Clauwaert P, Verstraete W. Microbial fuel cells for wastewater treatment. Water Sci Technol 2006; 54(8): 9-15.
[http://dx.doi.org/10.2166/wst.2006.702] [PMID: 17163008]

[27]   Ieropoulos IA, Greenman J, Melhuish C, Hart J. Comparative study of three types of microbial fuel cell. Enzyme Microb Technol 2005; 37(2): 238-45.
[http://dx.doi.org/10.1016/j.enzmictec.2005.03.006]

[28]   Chen GW, Choi SJ, Lee TH, Lee GY, Cha JH, Kim CW. Application of biocathode in microbial fuel cells: cell performance and microbial community. Appl Microbiol Biotechnol 2008; 79(3): 379-88.
[http://dx.doi.org/10.1007/s00253-008-1451-0] [PMID: 18385994]

[29]   Gorton L, Ed. Biosensors and modern biospecific analytical techniques. Elsevier 2005.

[30]   Ivask A, Green T, Polyak B, *et al.* Fibre-optic bacterial biosensors and their application for the analysis of bioavailable Hg and As in soils and sediments from Aznalcollar mining area in Spain. Biosens Bioelectron 2007; 22(7): 1396-402.
[http://dx.doi.org/10.1016/j.bios.2006.06.019] [PMID: 16889954]

[31]   Chinalia F, Paton G, Killham K. Physiological and toxicological characterization of an engineered whole-cell biosensor. Bioresour Technol 2008; 99(4): 714-21.
[http://dx.doi.org/10.1016/j.biortech.2007.01.041] [PMID: 17379508]

[32]   Girotti S, Ferri EN, Fumo MG, Maiolini E. Monitoring of environmental pollutants by bioluminescent bacteria. Anal Chim Acta 2008; 608(1): 2-29.
[http://dx.doi.org/10.1016/j.aca.2007.12.008] [PMID: 18206990]

[33]   Chee GJ, Nomura Y, Ikebukuro K, Karube I. Optical fiber biosensor for the determination of low biochemical oxygen demand. Biosens Bioelectron 2000; 15(7-8): 371-6.
[http://dx.doi.org/10.1016/S0956-5663(00)00093-2] [PMID: 11219750]

[34]   Plant TK, Chaplen FW, Jovanovic G, Kolodziej W, Trempy JE. Sensitive-cell-based fish chromatophore biosensor InBiomedical Vibrational Spectroscopy and Biohazard Detection Technologies. International Society for Optics and Photonics 2004; pp. 265-74.
[http://dx.doi.org/10.1117/12.528093]

[35]   Yoshida K, Yoshioka D, Inoue K, Takaichi S, Maeda I. Evaluation of colors in green mutants isolated from purple bacteria as a host for colorimetric whole-cell biosensors. Appl Microbiol Biotechnol 2007; 76(5): 1043-50.
[http://dx.doi.org/10.1007/s00253-007-1079-5] [PMID: 17609942]

[36]   Joel OF, Akinde BS, Nwokoye CU. Determination of Some Physicochemical and Microbiological Characteristics of Sewage Samples from Domestic, House-boat and Off-shore Facilities Discharge Point. J Appl Sci Environt Manage 2009; p. 13.

[37]   Musmade JB, Kakade JR, Pawar VM, Todmal VB. Biological oxygen demand and chemical oxygen demand content of mula dam, rahuri India, 2016.

[38]   Jouanneau S, Recoules L, Durand MJ, *et al.* Methods for assessing biochemical oxygen demand (BOD): A review. Water Res 2014; 49: 62-82.
[http://dx.doi.org/10.1016/j.watres.2013.10.066] [PMID: 24316182]

[39]   Monitoring of Indian National Aquatic Resources. Status of Water Quality in India 2011; 2013-4.

[40]   ENVIS Centre on Control of Pollution Water, Air and Noise. Hosted by Central Pollution Control Board.

[41]   State Pollution Control Boards/Pollution Control Committees under National Water Quality Monitoring Programme (NWMP), 2011.

[42]   Ponomareva ON, Arlyapov VA, Alferov VA, Reshetilov AN. Microbial biosensors for detection of biological oxygen demand (a Review). Appl Biochem Microbiol 2011; 47(1): 1-11.
[http://dx.doi.org/10.1134/S0003683811010108]

[43]   Omar KA. PREDICTION OF DISSOLVED OXYGEN IN TIGRIS RIVER BY WATER TEMPERATURE AND BIOLOGICAL OXYGEN DEMAND USING ARTIFICIAL NEURAL NETWORKS (ANNs). Journal of The University of Duhok 2017; 20(1): 791-00.

[http://dx.doi.org/10.26682/sjuod.2017.20.1.60]

[44]   Yoon Y, Kang Y, Chae Y, *et al.* Arsenic bioavailability in soils before and after soil washing: the use of *Escherichia coli* whole-cell bioreporters. Environ Sci Pollut Res Int 2016; 23(3): 2353-61.
[http://dx.doi.org/10.1007/s11356-015-5457-8] [PMID: 26411448]

[45]   Mohammadi MJ, Takdastan A, Jorfi S, *et al.* Electrocoagulation process to Chemical and Biological Oxygen Demand treatment from carwash grey water in Ahvaz megacity, Iran. Data Brief 2017; 11: 634-9.
[http://dx.doi.org/10.1016/j.dib.2017.03.006] [PMID: 28377993]

[46]   Logroño W, Guambo A, Pérez M, Kadier A, Recalde C. A terrestrial single chamber microbial fuel cell-based biosensor for biochemical oxygen demand of synthetic rice washed wastewater. Sensors (Basel) 2016; 16(1): 101.
[http://dx.doi.org/10.3390/s16010101] [PMID: 26784197]

[47]   Peixoto L, Min B, Martins G, *et al. In situ* microbial fuel cell-based biosensor for organic carbon. Bioelectrochemistry 2011; 81(2): 99-103.
[http://dx.doi.org/10.1016/j.bioelechem.2011.02.002] [PMID: 21371947]

[48]   Photphanloet C, Treeratanajaru W, Cooharojananone N, Lipikorn R. Biochemical oxygen demand prediction for Chaophraya river using alpha-trimmed ARIMA model. In 2016 13th International Joint Conference on Computer Science and Software Engineering (JCSSE) 2016; 1-6.
[http://dx.doi.org/10.1109/JCSSE.2016.7748930]

[49]   Salih SQ, Alakili I, Beyaztas U, Shahid S, Yaseen ZM. Prediction of dissolved oxygen, biochemical oxygen demand, and chemical oxygen demand using hydrometeorological variables. case study of Selangor River, Malaysia. Environ Dev Sustain 2020; 1-20.

[50]   Verma N, Singh AK. Development of biological oxygen demand biosensor for monitoring the fermentation industry effluent. Int Sch Res Notices 2013.
[PMID: 25969770]

[51]   Axelrod T, Eltzov E, Marks RS. Bioluminescent bioreporter pad biosensor for monitoring water toxicity. Talanta 2016; 149: 290-7.
[http://dx.doi.org/10.1016/j.talanta.2015.11.067] [PMID: 26717844]

[52]   Pirofski LA, Casadevall AQ. Q&A: What is a pathogen? A question that begs the point. J BMC Bio 2012; 10: 6.

[53]   Kumar N, Hu Y, Singh S, Mizaikoff B. Emerging biosensor platforms for the assessment of water-borne pathogens. Analyst (Lond) 2018; 143(2): 359-73.
[http://dx.doi.org/10.1039/C7AN00983F] [PMID: 29271425]

[54]   Hunter PR. Climate change and waterborne and vector-borne disease. J Appl Microbiol 2003; 94 (Suppl.): 37-46.
[http://dx.doi.org/10.1046/j.1365-2672.94.s1.5.x] [PMID: 12675935]

[55]   Justino C, Duarte A, Rocha-Santos T. Recent progress in biosensors for environmental monitoring. A review. Sensors (Basel) 2017; 17(12): 2918.
[http://dx.doi.org/10.3390/s17122918] [PMID: 29244756]

[56]   Xu X, Ying Y. Microbial biosensors for environmental monitoring and food analysis. Food Rev Int 2011; 27(3): 300-29.
[http://dx.doi.org/10.1080/87559129.2011.563393]

[57]   Upadhyayula VK, Ghoshroy S, Nair VS, Smith GB, Mitchell MC. Single-walled carbon nanotubes as fluorescence biosensors for pathogen recognition in water systems. J Nanotechnol 2008.

[58]   Zhan F, Wang T, Iradukunda L, Zhan J. A gold nanoparticle-based lateral flow biosensor for sensitive visual detection of the potato late blight pathogen, Phytophthora infestans. Anal Chim Acta 2018; 1036: 153-61.
[http://dx.doi.org/10.1016/j.aca.2018.06.083] [PMID: 30253826]

[59]   Valadez A, Lana C, Tu SI, Morgan M, Bhunia A. Evanescent wave fiber optic biosensor for salmonella detection in food. Sensors (Basel) 2009; 9(7): 5810-24.
[http://dx.doi.org/10.3390/s90705810] [PMID: 22346728]

[60]   Khan M, Khan AR, Shin JH, Park SY. A liquid-crystal-based DNA biosensor for pathogen detection. Sci Rep 2016; 6(1): 22676.
[http://dx.doi.org/10.1038/srep22676] [PMID: 26940532]

[61]   Dutse SW, Yusof NA, Ahmad H, Hussein MZ, Hushiarian R. DNA-based biosensor for detection of Ganodermaboninense, an Oil palm pathogen utilizing newly synthesized ruthenium complex [Ru (phen) 2 (qtpy)] 2 based on a PEDOT-PSS/Ag nanoparticle modified electrode. Int J Electrochem Sci 2013; 8: 11048-57.

[62]   Mukundan H, Kumar S, Price DN, *et al.* Rapid detection of Mycobacterium tuberculosis biomarkers in a sandwich immunoassay format using a waveguide-based optical biosensor. Tuberculosis (Edinb) 2012; 92(5): 407-16.
[http://dx.doi.org/10.1016/j.tube.2012.05.009] [PMID: 22710249]

[63]   Wang L, Wei Q, Wu C, Hu Z, Ji J, Wang P. The *Escherichia coli* O157:H7 DNA detection on a gold nanoparticle-enhanced piezoelectric biosensor. Sci Bull (Beijing) 2008; 53(8): 1175-84.
[http://dx.doi.org/10.1007/s11434-007-0529-x]

[64]   Hashemi Tameh M. Pectobacteriumatrosepticum biosensor for monitoring blackleg and soft rot disease of potato. J Biosci 2020; 10: 64.

[65]   Ostrov N, Jimenez M, Billerbeck S, *et al.* A modular yeast biosensor for low-cost point-of-care pathogen detection. Sci Adv 2017; 3(6): e1603221.
[http://dx.doi.org/10.1126/sciadv.1603221] [PMID: 28782007]

[66]   Cho IH, Irudayaraj J. *in situ* immuno-gold nanoparticle network ELISA biosensors for pathogen detection. Int J Food Microbiol 2013; 164(1): 70-5.
[http://dx.doi.org/10.1016/j.ijfoodmicro.2013.02.025] [PMID: 23603219]

[67]   Kim G, Morgan M, Hahm BK, Bhunia A, Mun JH, Om AS. Interdigitated microelectrode-based impedance biosensor for detection of Salmonella enteritidis in food samples. J Phy: Conference Series 2008; 100: 5-052044.
[http://dx.doi.org/10.1088/1742-6596/100/5/052044]

[68]   Khan MZH, Hasan MR, Hossain SI, Ahommed MS, Daizy M. Ultrasensitive detection of pathogenic viruses with electrochemical biosensor: State of the art. Biosens Bioelectron 2020; 166: 112431.
[http://dx.doi.org/10.1016/j.bios.2020.112431] [PMID: 32862842]

[69]   Vizzini P, Braidot M, Vidic J, Manzano M. Electrochemical and optical biosensors for the detection of campylobacter and listeria. An update look. Micromachines (Basel) 2019; 10(8): 500.
[http://dx.doi.org/10.3390/mi10080500] [PMID: 31357655]

[70]   Li S, Li Y, Chen H, *et al.* Direct detection of Salmonella typhimurium on fresh produce using phage-based magnetoelastic biosensors. Biosens Bioelectron 2010; 26(4): 1313-9.
[http://dx.doi.org/10.1016/j.bios.2010.07.029] [PMID: 20688505]

[71]   Vikesland PJ, Wigginton KR. Nanomaterial enabled biosensors for pathogen monitoring - a review. Environ Sci Technol 2010; 44(10): 3656-69.
[http://dx.doi.org/10.1021/es903704z] [PMID: 20405829]

[72]   Koedrith P, Thasiphu T, Weon JI, Boonprasert R, Tuitemwong K, Tuitemwong P. Recent trends in rapid environmental monitoring of pathogens and toxicants: potential of nanoparticle-based biosensor and applications. Sci World J 2015.

[73]   Wang Y, Ye Z, Ying Y. New trends in impedimetric biosensors for the detection of foodborne pathogenic bacteria. Sensors (Basel) 2012; 12(3): 3449-71.
[http://dx.doi.org/10.3390/s120303449] [PMID: 22737018]

[74]   Kumar A, Grover S, Batish VK. Monitoring paneer for Listeria monocytogenes-A high risk food pathogen by multiplex PCR. Afr J Biotechnol 2012; 11: 9452-6.

[75]   Poltronieri P, Mezzolla V, Primiceri E, Maruccio G. Biosensors for the detection of food pathogens. Foods 2014; 3(3): 511-26.
[http://dx.doi.org/10.3390/foods3030511] [PMID: 28234334]

[76]   Liu F, Choi KS, Park TJ, Lee SY, Seo TS. Graphene-based electrochemical biosensor for pathogenic virus detection. Biochip J 2011; 5(2): 123-8.
[http://dx.doi.org/10.1007/s13206-011-5204-2]

[77]   Wang Y, Ye Z, Si C, Ying Y. Monitoring of *Escherichia coli* O157:H7 in food samples using lectin based surface plasmon resonance biosensor. Food Chem 2013; 136(3-4): 1303-8.
[http://dx.doi.org/10.1016/j.foodchem.2012.09.069] [PMID: 23194527]

[78]   Hu L, Ge A, Wang X, *et al.* Real-time monitoring of immune responses under pathogen invasion and drug interference by integrated microfluidic device coupled with worm-based biosensor. Biosens Bioelectron 2018; 110: 233-8.
[http://dx.doi.org/10.1016/j.bios.2018.03.058] [PMID: 29625331]

[79]   Vinayaka AC, Thakur MS. Focus on quantum dots as potential fluorescent probes for monitoring food toxicants and foodborne pathogens. Anal Bioanal Chem 2010; 397(4): 1445-55.
[http://dx.doi.org/10.1007/s00216-010-3683-y] [PMID: 20401471]

[80]   Bereza-Malcolm LT, Mann G, Franks AE. Environmental sensing of heavy metals through whole cell microbial biosensors: a synthetic biology approach. ACS Synth Biol 2015; 4(5): 535-46.
[http://dx.doi.org/10.1021/sb500286r] [PMID: 25299321]

[81]   Gupta N, Renugopalakrishnan V, Liepmann D, Paulmurugan R, Malhotra BD. Cell-based biosensors: Recent trends, challenges and future perspectives. Biosens Bioelectron 2019; 141: 111435.
[http://dx.doi.org/10.1016/j.bios.2019.111435] [PMID: 31238280]

[82]   Guo M, Wang J, Du R, *et al.* A test strip platform based on a whole-cell microbial biosensor for simultaneous on-site detection of total inorganic mercury pollutants in cosmetics without the need for predigestion. Biosens Bioelectron 2020; 150: 111899.
[http://dx.doi.org/10.1016/j.bios.2019.111899] [PMID: 31767350]

[83]   Pepi M, Reniero D, Baldi F, Barbieri P. A comparison of mer:: lux whole cell biosensors and moss, a bioindicator, for estimating mercury pollution. J Water, Air, and Soil Pollution 2006; 173: 163-75.

[84]   Chen J, Rosen B. Biosensors for inorganic and organic arsenicals. Biosensors (Basel) 2014; 4(4): 494-512.
[http://dx.doi.org/10.3390/bios4040494] [PMID: 25587436]

[85]   Chen SY, Wei W, Yin BC, Tong Y, Lu J, Ye BC. Development of a highly sensitive whole-cell biosensor for arsenite detection through engineered promoter modifications. J. ACS Synth Biol 2019; 8(10): 2295-302.
[http://dx.doi.org/10.1021/acssynbio.9b00093] [PMID: 31525958]

[86]   Jang YC, Somanna Y, Kim H. Source, distribution, toxicity and remediation of arsenic in the environment–a review. Int J Appl Environt Sci 2016; 11: 559-81.

[87]   Joe MH, Lee KH, Lim SY, *et al.* Pigment-based whole-cell biosensor system for cadmium detection using genetically engineered Deinococcus radiodurans. Bioprocess Biosyst Eng 2012; 35(1-2): 265-72.
[http://dx.doi.org/10.1007/s00449-011-0610-3] [PMID: 21928095]

[88]   Elcin E, Öktem HA. Inorganic Cadmium Detection Using a Fluorescent Whole-Cell Bacterial Bioreporter. Anal Lett 2020; 53(17): 2715-33.
[http://dx.doi.org/10.1080/00032719.2020.1755867]

[89]   Bereza-Malcolm L, Aracic S, Franks A. Development and application of a synthetically-derived lead biosensor construct for use in gram-negative bacteria. Sensors (Basel) 2016; 16(12): 2174.

[http://dx.doi.org/10.3390/s16122174] [PMID: 27999352]

[90]   Lee JI, Karube I. A novel microbial sensor for the determination of cyanide. Anal Chim Acta 1995; 313(1-2): 69-74.
[http://dx.doi.org/10.1016/0003-2670(95)00232-O]

[91]   Nakanishi K, Ikebukuro K, Karube I. Determination of cyanide using a microbial sensor. Appl Biochem Biotechnol 1996; 60(2): 97-106.
[http://dx.doi.org/10.1007/BF02788064] [PMID: 8856940]

[92]   Mak KKW, Law AWC, Tokuda S, Yanase H, Renneberg R. Application of cyanide hydrolase from Klebsiella sp. in a biosensor system for the detection of low-level cyanide. Appl Microbiol Biotechnol 2005; 67(5): 631-6.
[http://dx.doi.org/10.1007/s00253-004-1825-x] [PMID: 15630582]

[93]   Virender Kumar , Kumar V, Singh AK, Verma N, Bhalla TC. A potentiometric biosensor for cyanide detection using immobilized whole cell cyanide dihydratase of Flavobacteriumindicum MTCC 6936. J Anal Chem 2018; 73(10): 1014-9.
[http://dx.doi.org/10.1134/S1061934818100039]

[94]   Stenzler BR, Gaudet J, Poulain AJ. An anaerobic biosensor assay for the detection of mercury and cadmium. J Vis Exp 2018; 142(142): 58324.
[http://dx.doi.org/10.3791/58324] [PMID: 30596384]

[95]   Stenzler B, Hinz A, Ruuskanen M, Poulain AJ. Ionic strength differentially affects the bioavailability of neutral and negatively charged inorganic Hg complexes. Environ Sci Technol 2017; 51(17): 9653-62.
[http://dx.doi.org/10.1021/acs.est.7b01414] [PMID: 28701033]

[96]   Sciuto EL, Coniglio MA, Corso D, *et al.* Biosensors in Monitoring Water Quality and Safety: An Example of a Miniaturizable Whole-Cell Based Sensor for Hg^{2+} Optical Detection in Water. Water 2019; 11(10): 1986.
[http://dx.doi.org/10.3390/w11101986]

[97]   Roointan A, Shabab N, Karimi J, Rahmani A, Alikhani MY, Saidijam M. Designing a bacterial biosensor for detection of mercury in water solutions. Turk J Biol 2015; 39: 550-5.
[http://dx.doi.org/10.3906/biy-1411-49]

[98]   Cai S, Shen Y, Zou Y, *et al.* Engineering highly sensitive whole-cell mercury biosensors based on positive feedback loops from quorum-sensing systems. Analyst (Lond) 2018; 143(3): 630-4.
[http://dx.doi.org/10.1039/C7AN00587C] [PMID: 29271434]

[99]   Bontidean I, Mortari A, Leth S, *et al.* Biosensors for detection of mercury in contaminated soils. Environ Pollut 2004; 131(2): 255-62.
[http://dx.doi.org/10.1016/j.envpol.2004.02.019] [PMID: 15234092]

[100]   Jia X, Bu R, Zhao T, Wu K. Sensitive and specific whole-cell biosensor for arsenic detection. Appl Environ Microbiol 2019; 85(11): e00694-19.
[http://dx.doi.org/10.1128/AEM.00694-19] [PMID: 30952659]

[101]   Pothier MP, Hinz AJ, Poulain AJ. Insights into arsenite and arsenate uptake pathways using a whole cell biosensor. Front Microbiol 2018; 9: 2310.
[http://dx.doi.org/10.3389/fmicb.2018.02310] [PMID: 30333804]

[102]   Elcin E, Öktem HA. Whole-cell fluorescent bacterial bioreporter for arsenic detection in water. Int J Environ Sci Technol 2019; 16(10): 5489-500.
[http://dx.doi.org/10.1007/s13762-018-2077-0]

[103]   Dieudonne A, Preveral S, Pignol D. A sensitive magnetic arsenite-specific biosensor hosted in magnetotactic bacteria. J Appl Environt Microbiol 2020.
[http://dx.doi.org/10.1128/AEM.00803-20]

[104]   Hurdebise Q, Tarayre C, Fischer C, Colinet G, Hiligsmann S, Delvigne F. Determination of zinc,

cadmium and lead bioavailability in contaminated soils at the single-cell level by a combination of whole-cell biosensors and flow cytometry. Sensors (Basel) 2015; 15(4): 8981-99.
[http://dx.doi.org/10.3390/s150408981] [PMID: 25894939]

[105]   Tra-ngan S, Siripornadulsil S, Thanwisai L, Siripornadulsil W. Potential application of a recombinant bacterial strain carrying a groEL promoter as a whole-cell microbial biosensor for detecting bioavailable cadmium. Environmental Technology & Innovation 2019; 15: 100375.
[http://dx.doi.org/10.1016/j.eti.2019.100375]

[106]   Ratajczak A, Geißdörfer W, Hillen W. Alkane hydroxylase from Acinetobacter sp. strain ADP1 is encoded by alkM and belongs to a new family of bacterial integral-membrane hydrocarbon hydroxylases. Appl Environ Microbiol 1998; 64(4): 1175-9.
[http://dx.doi.org/10.1128/AEM.64.4.1175-1179.1998] [PMID: 9546151]

[107]   Tauriainen S, Karp M, Chang W, Virta M. Luminescent bacterial sensor for cadmium and lead. Biosens Bioelectron 1998; 13(9): 931-8.
[http://dx.doi.org/10.1016/S0956-5663(98)00027-X] [PMID: 9839381]

[108]   Hui C, Guo Y, Liu L, *et al.* Genetic control of violacein biosynthesis to enable a pigment-based whole-cell lead biosensor. RSC Advances 2020; 10(47): 28106-13.
[http://dx.doi.org/10.1039/D0RA04815A] [PMID: 35519119]

[109]   Chakraborty T, Babu PG, Alam A, Chaudhari A. GFP expressing bacterial biosensor to measure lead contamination in aquatic environment. J Curr Sci 2008; 25: 800-5.

[110]   Okochi M, Mima K, Miyata M, *et al.* Development of an automated water toxicity biosensor usingThiobacillus ferrooxidans for monitoring cyanides in natural water for a water filtering plant. Biotechnol Bioeng 2004; 87(7): 905-11.
[http://dx.doi.org/10.1002/bit.20193] [PMID: 15334417]

CHAPTER 7

Microbial Degradation, Bioremediation and Biotransformation

Sameena Lone[1,*], **Khursheed Hussain**[1] and **Abdel Rahman Al-Tawaha**[2]

[1] Sher-e-Kashmir University of Agricultural Sciences and Technology, Kashmir, Shalimar, 190 025

[2] Department of Biological Sciences, Al-Hussein bin Talal University, Maan, Jordan

Abstract: At present, the world is reeling under the problem of different environmental pollutions, *viz.*, soil, water, and air pollution, as a result of anthropogenic activities, intensive inorganic agriculture, industrial revolution releasing a wide array of xenobiotics. Across the world, scientists are trying to overcome pollution through physical, chemical, and thermal processes. The major drawbacks of these methods include their labor-intensive nature, high cost, and undesirable changes in the treated soil's physical, chemical and biological characteristics. The only alternative solution to overcome this challenge is microorganisms. The microorganisms transform the various substances through their metabolic activity. It mainly depends on two processes. growth and cometabolism. Growth refers to the process which results in complete degradation (mineralization) of organic pollutants. Hence, the only source of carbon and energy in growth is an organic pollutant.

On the other hand, cometabolism refers to the process in which the metabolism of an organic compound takes place in the presence of a growth substrate, which is used as the primary source of carbon and energy. For maintaining the global carbon cycle and renewing our environment, microorganisms have an essential role to play. The various microbial activities are comprehended in biodegradation, bioremediation, and biotransformation. Substances transformed by microorganisms include a wide range of synthetic compounds and other chemical substances like hydrocarbons and heavy metals, which have toxic ecological effects. However, in most cases, this statement is concerned with the potential degradabilities of microorganisms estimated under ideal growth conditions using selected laboratory cultures.

Keywords: Biodegradation, Bioremediation, Biotransformation Cometabolism, Microorganisms.

---

[*] **Corresponding author Sameena Lone:** Sher-e-Kashmir University of Agricultural Sciences and Technology, Kashmir, Shalimar, India - 190025; E-mails: sameenalone77@skuastkashmir.ac.in

Arun Karnwal & Abdel Rahman Mohammad Said Al-Tawaha (Eds.)

INTRODUCTION

Currently, the world is reeling under the problem of different environmental pollution, *viz.*, soil, water, and air pollution, as a result of anthropogenic activities, intensive inorganic agriculture, industrial revolution releasing a wide array of xenobiotics. Across the world, scientists are trying to overcome pollution through physical, chemical, and thermal processes. The major drawbacks of these methods include their labor-intensive nature, high cost, and undesirable changes in the treated soil's physical, chemical and biological characteristics. Microorganisms are the sole option for overcoming this problem. They survive everywhere in this biosphere because of the astonishing metabolic activity that enables them to survive under various environmental conditions. They vary in their nutritional requirements, *i.e.*, why they are used as bioremediation measures to tackle the various kinds of environmental pollutants, as bioremediation is highly involved in the degradation, eradication, immobilization, or detoxification of various chemical wastes and physical hazardous materials from the surroundings through the activity of microorganisms [1 - 5]. The main principle on which biodegradation depends is to degrade and transform pollutants such as hydrocarbons, oils, heavy metals, pesticides, dyes, *etc.*, which is carried out in an enzymatic way by metabolizing these pollutants. Hence, it plays a vital role in solving many environmental issues. The degradation rate is usually determined by two factors *i.e.*, biotic and abiotic factors. However, enormous methods and strategies have been applied in different parts of the world to solve these environmental concerns. The microorganisms transform the various substances through their metabolic activity [3]. This transformation mainly depends on two processes. growth and cometabolism. Growth refers to the process which results in complete degradation (mineralization) of organic pollutants. Hence, the only source of carbon and energy in growth is an organic pollutant. On the other hand, cometabolism refers to the process in which the metabolism of an organic compound takes place in the presence of a growth substrate, which is used as the primary source of carbon and energy [1].

In biodegradation, various microorganisms are involved, including fungi, bacteria, and yeasts. The biodegradation processes differ significantly, but carbon dioxide is always the final result [5].

Microbial Biodegradation

Biodegradation is characterized as the biologically catalyzed decrease in chemical complexity [6]. In other words, the method of biodegradation is the division of organic substances by living microorganisms into more minor compounds. Once biodegradation is complete, mineralization follows [3].

Types of Biodegradation

Aerobic Biodegradation

It is defined as the process in which the degradation of organic pollutants, *i.e.*, the organic decomposition, takes place under aerobic conditions (in the presence of oxygen) and is done by aerobic bacteria.

Anaerobic Biodegradation

In this process, organic decomposition takes place only in the absence of oxygen. In the anaerobic biodegradation process, the four main stages are hydrolysis, acidogenesis, acetogenesis, and methane production. All organic substrates are neutralized by anaerobic digestion to produce carbon dioxide, which is converted into methane [7].

Role of Microbes in Biodegradation

The process of biodegradation can be divided into three stages [3]:

Biodeterioration- means degradation at a surface level which results in the alteration of mechanical, physical, and chemical properties of the target material by exposing the target material to some abiotic factors (mechanical compression, light, temperature, or even chemicals) present in the environment which weakens the structure of the material [6].

Biofragmentation- is the process in which a polymer is fragmented or broken down into its constituent oligomers and monomers, and this breakdown of material by microorganisms may occur either in the presence of oxygen (aerobic bio-fragmentation) without the production of methane or in the absence of oxygen (anaerobic bio-fragmentation) with the production of methane. However, the production of CO_2, H_2O, some residues, and a new biomass is a feature of both subtypes. There is a great advantage associated with aerobic digestion, which is its fast speed than anaerobic digestion; however, the latter has the advantage of reducing the volume and mass of the target material with the production of natural gas. This is why anaerobic bio-fragementation technology finds its application in converting wastes into useful ones, aiding in the process of renewable energy sources.

The Assimilation Stage: is where the end products of the bio-fragmentation stage are integrated into the microbial cells. Few products from the fragmentation stage get easily transported within the cell with the help of carriers present in the cell membranes; however, the products which are large enough to go inside the cells need to be further bio-fragmented (aerobic or anaerobically) to form smaller substances that can quickly go inside the cells of microbes so that they are assimilated. In this way, the bio-fragmentation makes the molecules available for the microbial cells for further catabolism resulting in the production of universal energy currency Adenosine Tri- Phosphate (ATP) or elements of the structural components of the cell [4].

A single microbe cannot degrade all the organic pollutants present in the environment into simpler forms, but different microbes possess the ability to convert different pollutants; therefore, mixed microbial communities with different bio-fragmentation potentials are required to degrade the complex mixtures of organic compounds present in contaminated areas which results in the bacterial consortium formulation development [22, 23]. Several evidences have been reported on the degradation of different environmental pollutants by different bacteria that are even known to feed exclusively on hydrocarbons. Bacteria that can cause hydrocarbon degradation are known as hydrocarbon-degrading bacteria. Biodegradation of hydrocarbons can also occur under aerobic or anaerobic conditions [5, 9].

About 80 bacterial strains belonging to 10 genera (*Bacillus, Corynebacterium, Staphylococcus, Streptococcus, Shigella, Alcaligenes, Acinetobacter, Escherichia, Klebsiella,* and *Enterobacter*) were isolated [2]. *Bacillus* ranked as the best hydrocarbon-degrading bacteria. Reports confirm that at least 25 genera of hydrocarbon-degrading bacteria have been isolated from marine environments [2]. Bio-degradation pathways have been reported in bacteria from Mycobacterium, Corynebacterium, Aeromonas, Rhodococcus, and Bacillus. The bacterial strains capable of degrading hydrocarbons have been isolated from soil; most belong to the genus *Pseudomonas*. This aerobic or anaerobic degradation occurs in the case of nitrate-reducing bacterial strains such as *Pseudomonas* sp. and *Brevibacillus* sp., which are isolated from soils contaminated with petroleum [10]. It is reported that anaerobic biodegradation may be much more important.

Microbial Bioremediation

Bioremediation is a natural biological process rather than a physical or chemical process whereby organic pollutants are biologically degraded into simpler products that are less harmful and harmless either in nature, or their levels are reduced below the harmful levels under conditions that are controlled with the help of living organisms, most notably microorganisms such as bacteria, fungi, or plants, or their derivatives such as enzymes due to their physiological competence [11]. The microorganisms can be present on the site (indigenous microorganisms) or isolated from other sites and added to the treated material. This process has multifarious advantages over physico-chemical methods as it is cost-effective, versatile, efficient, and eco-friendly [12, 13]. However, for microbial culture and biomass enhancement, well-designed bioreactors with controlled environmental conditions are to be used [5].

Types of Bioremediation

Biostimulation- is the process in which the surrounding environment is changed or modified in such a way that existing indigenous or naturally occurring bacteria and fungus communities are stimulated to ensure bioremediation by adding different types of specific nutrients which are limited in soil and include macronutrients (as N, P, K) or micronutrients (as Mg, S, Fe, Cl, Zn, Mn, Cu, Na) in the form of molasses, fertilizers, growth supplements, and trace minerals, changing pH, soil temperature and oxygen to speed up their metabolism rate and pathway), important to recover depleted soils by agricultural management systems. It has been concluded that specific nutrients create basic building blocks of life (nitrogen, phosphorous, and carbon), provide energy, increase cell biomass and form enzymes required for the degradation of the pollutants.

Bioattenuation- is the natural eradication of a pollutant or contaminant through biological transformation, through which the natural microbial populations degrade the recalcitrant or xenobiotic compounds *via* their metabolic processes by various chemical, physical and biological processes that reduce the mass, volume or toxicity, or concentration. Among forty polycyclic aromatic hydrocarbons (PAH)-degrading bacteria, *Cycloclasticus* and *Pseudomonas* exhibited the maximum ability to degrade PAH under low temperatures. The other three genera *viz., Pseudoalteromonas, Halomonas, Marinomonas* and *Dietzia* are important for PAH mineralization *in situ*. However, if the natural attenuation is slow or incomplete, bioremediation will be enhanced either by biostimulation or bioaugmentation.

Bioaugmentation - is the process of adding cultured microorganisms into the soil subsurface to augment the biodegradation capacity of local microbial populations so that the pollutants get degraded in soil and groundwater. This is a simple process of collecting the microbes from the remediation place, then culturing and modifying genetically microbes, which completely degrade complex pollutants [14]. The remediation places are where we have already contaminated soil or groundwater with chlorinated ethene (tetrachloroethylene and trichloroethylene) because these *in situ* microbes can remove and alter these contaminants to ethylene and chloride, which are non-toxic. It has been reported that genetically modified microbes increase the degradative efficiency of various environmental pollutants due to their diverse metabolic pathways to convert complex pollutants to less complex and harmless end products [15]. In contrast, wild microbes are not quick to break down or degrade pollutants and show high competition with indigenous species and predators besides resilience to various abiotic factors. Thus bioremediation of soil, groundwater, and activated sludge is a simple problem in the future.

Bioventing- oxygen is deficient in soil (unsaturated soil), which is a reason for the naturally occurring microbial populations to have slow growth and thus a sluggish biodegradation rate. Therefore, by venting enough oxygen through low airflow rates to provide enough oxygen to sustain microbial activity in the soil, stimulating the natural *in situ* biodegradation of soil pollutants is bioventing. Oxygen may be augmented through direct air injection into the soil (biologically active soil) *via* bioventing wells or vent well. On a vent well is installed a valve (one way) through which air enters the well only when the average air pressure inside the well is lower than atmospheric pressure. However, if atmospheric pressure falls less than subsurface air pressure, the valve closes, trapping the air in the well and increasing oxygen to the soil surrounding the well. Some problems are associated with the bioventing phenomenon, like the presence of high soil moisture; low soil temperature decreases air permeability into the otherwise unsaturated soil reducing bioventing performance. This technology has been commercially used to remediate soils that are polluted with non-chlorinated (because aerobic biodegradation of a large number of chlorinated compounds is not so effective unless a cometabolite (enzyme from microbiological metabolism causing degradation of a pollutant or a contaminant) is present), some pesticides, wood preservatives, and other organic chemicals. Cometabolism is a process whereby a dilute solution of nutrients such as methane and oxygen is injected into the soil or groundwater polluted so that the microbes utilizing these nutrients metabolize to produce enzymes that react with the organic contaminant cause its degradation to harmless minerals. For example, adding methane or methanol degrades chlorinated solvents, such as vinyl chloride and trichloroethylene.

Biopiles- an *ex-situ* technology in which the soil is contaminated with aerobically remediable hydrocarbons is excavated. Then this excavated soil is added with soil amendments, later made into soil piles or compost piles (can be up to or even more than 6 m high though ideal height is 2-3 meters), enclosed for treatment simultaneously stimulating the biodegrading activity of microorganisms by allowing optimum growth conditions by using biological systems (microorganisms, plants, or enzymes) [16 - 18]. The biocells, bioheaps, biomounds, and compost piles are also used to decrease petroleum pollutant concentrations in excavated soils during biodegradation. During this process, the air is supplied to the biopillar device, which either pushes air into the pile under positive pressure or draws air through the pile under negative pressure [18]. The microbial function is stimulated by microbial respiration, resulting in the high degradation of adsorbed petroleum contaminants [17].

Role of Microbes in Bioremediation

In simpler terms, microbes that live in soil and groundwater like to eat certain harmful chemicals, and when microbes completely digest these chemicals, they change them into the water and harmless gases such as carbon dioxide using natural biological activity. Microorganisms ensure contaminant destruction due to the presence of enzymes, which microbes use as environmental contaminants as their food. All metabolic reactions are mediated by these enzymes, which may be among the groups of oxidoreductases, hydrolases, lyases, transferases, isomerases, and ligases. The metabolic machinery can degrade most of the synthetic compounds such as hydrocarbons (*e.g.*, oil), polychlorinated biphenyls (PCBs), polyaromatic hydrocarbons (PAHs), radionuclides and metals. As far heavy metals are concerned, these cannot be destroyed biologically or cannot be degraded further, but such heavy metals can be transformed only from one oxidation state or organic complex to other states [19, 20] by various mechanisms, such as adsorption, uptake, methylation, oxidation, and reduction. Microbes' uptake of these heavy metals is known as bioaccumulation and/or passively known as adsorption. It is important to note that methylation by microbes for heavy metals is a boon because methylated compounds are frequently volatile. Biomethylation can occur by several different bacterial species such as *Alcaligenes faecalis*, *Bacillus pumilus*, *Bacillus sp.*, *P. aeruginosa,* and *Brevibacterium iodinium* for instance mercury, Hg (II) can be changed to gaseous methyl mercury. Bioremediation is very useful from the human health point of view. Though metals are essential to the metabolic functions of plants and animals, at increased levels, they interfere with metabolic reactions in the physiological systems of organisms. Few toxic heavy metals *viz.*, lead (Pb), cadmium (Cd), mercury (Hg), chromium (Cr), zinc (Zn), uranium (Ur), selenium (Se), silver (Ag), gold (Au), nickel (Ni) and arsenic (As) are not important to plants but reduce plant growth because of reduced photosynthetic activities, plant mineral nutrition, and activity of essential enzymes, cause cytotoxicity at low concentrations and may even cause cancer in humans [21, 22]. Heavy metals cause toxicity and lead to the production of reactive oxygen species (ROS) hence decreasing the antioxidant systems (glutathione, superoxide dismutase, *etc.*), which protect cells [23, 24].

Microbes have a role in bioremediation in optimum and extreme environmental conditions because they are adapted to subzero temperatures, extreme heat, desert conditions, waterlogged conditions, excess oxygen, and an absence of oxygen, in the presence of hazardous compounds or any waste area. Examples of some bacteria that actively participate in bioremediation are as under [23, 24].

Aerobic- Microbes that thrive well in the presence of oxygen. They help in the degradation of pesticides and hydrocarbons [27], alkanes and polyaromatic compounds. Most of these bacteria use the contaminant as their primary carbon and energy source. These bacteria include Pseudomonas, Alcaligenes, Sphingomonas, Rhodococcus, and Mycobacterium [28].

Anaerobic- Microbes that do not thrive well in the presence of oxygen. They degrade polychlorinated biphenyls (PCBs), resulting in the dechlorination of the solvent trichloroethylene (TCE) and chloroform.

Methylotrophs – a kind of aerobic bacteria that grow by utilizing methane for carbon and energy. The initial mechanism for aerobic microbial degradation of chlorinated aromatic hydrocarbons includes methane monooxygenase, which is active against a wide variety of compounds, including, for example, trichloroethylene and 1,2-dichloroethane.

The benefits of microbial bioremediation are that the public recognizes it as a natural process. In most cases, low-cost equipment is compared to other methods of cleaning hazardous waste. It can be achieved *in-situ* and *ex-situ*; instead of moving contaminants from one type to another or from one medium to another, destruction of target organic pollutants is possible. Notable drawbacks are that bioremediation takes a relatively long time to achieve treatment goals, that it may not be successful for all contaminants, that certain biodegradation products may be more harmful or persistent than the parent compound, that the specificity of biological processes concerning microbial species, pollutants and environmental constraints is also a drawback and that specialized expertise is accessible.

Table 1. Bioremediation of specific pollutants using various microorganisms [7, 15, 22, 28, 29].

S. No.	Microbes	Pollutant
Hydrocarbon Bioremediation		
1.	*Penicillium chrysogenum*	Monocyclic aromatic hydro carbons (MAH), benzene, toluene, ethyl benzene and xylene, phenol compounds
2.	*P. alcaligenes P. mendocina and P. putida P. veronii, Achromobacter, Flavobacterium, Acinetobacter*	Petrol and diesel polycyclic aromatic hydrocarbons toluene
3.	*Pseudomonas putida*	Monocyclic aromatic hydrocarbons, *e.g.* benzene and xylene.
4.	*Phanerochaete chrysosporium*	Biphenyl and triphenylmethane
5.	*A. niger, A. fumigatus, F. solani and P. funiculosum*	Hydrocarbon

(Table 1) cont.....

S. No.	Microbes	Pollutant
6.	*Coprinellus radians*	PAHs, methylnaphthalenes, and dibenzofurans
7.	*Alcaligenes odorans, Bacillus subtilis, Corynebacterium propinquum, Pseudomonas aeruginosa*	Phenol
8.	*Tyromyces palustris, Gloeophyllum trabeum, Trametes versicolor*	Hydrocarbons
9.	*Candida viswanathii*	Phenanthrene, benzopyrene
10.	*cyanobacteria, green algae and diatoms and Bacillus licheniformis*	Naphthalene
11.	*Acinetobacter sp., Pseudomonas sp., Ralstonia sp. and Microbacterium sp,*	aromatic hydrocarbons
12.	*Gleophyllum striatum*	Striatum Pyrene, anthracene, 9-metil anthracene, Dibenzothiophene Lignin peroxidasse
13.	*Acinetobacter sp., Pseudomonas sp., Ralstonia sp. and Microbacterium sp,*	aromatic hydrocarbons
14.	*Gleophyllum striatum*	Striatum Pyrene, anthracene, 9-metil anthracene, Dibenzothiophene Lignin peroxidasse
15.	*Acinetobacter sp., Pseudomonas sp., Ralstonia sp. and Microbacterium sp,*	Aromatic hydrocarbons
16.	*Gleophyllum striatum*	Striatum Pyrene, anthracene, 9-metil anthracene, Dibenzothiophene Lignin peroxidase
Oil Bioremediation		
17.	*Fusarium* sp.	Oil
18.	*Alcaligenes odorans, Bacillus subtilis, Corynebacterium propinquum, Pseudomonas aeruginosa*	Oil
19.	*Bacillus cereus* A	Diesel oil
20.	*Aspergillus niger, Candida glabrata, Candida krusei and Saccharomyces cerevisiae*	Crude oil
21.	*B. brevis, P. aeruginosa KH6, B. licheniformis* and *B. sphaericus*	Crude oil
22.	*Pseudomonas aeruginosa, P. putida, Arthobacter sp and Bacillus sp*	Diesel oil
23.	*Pseudomonas cepacia, Bacillus cereus, Bacillus coagulans, Citrobacter koseri and Serratia ficaria*	Diesel oil, crude oil
	Representative examples of most dominate microorganisms in the involvement of dyes bioremadation.	

(Table 1) cont.....

S. No.	Microbes	Pollutant
24.	*B. subtilis* strain NAP1, NAP2, NAP4	Oil-based based paints
25.	*Myrothecium roridum* IM 6482	Industrial dyes
26.	*Pycnoporus sanguineous, Phanerochaete chrysosporium and Trametes trogii*	Industrial dyes
27.	*Penicillium ochrochloron*	Industrial dyes
28.	*Micrococcus luteus, Listeria denitrificans and Nocardia atlantica*	Textile Azo Dyes
29.	*Bacillus spp.* ETL-2012, *Pseudomonas aeruginosa, Bacillus pumilus* HKG212	Textile Dye (Remazol Black B), Sulfonated di-azo dye Reactive Red HE8B, RNB dye
30.	*Exiguobacterium indicum, Exiguobacterium aurantiacums, Bacillus cereus and Acinetobacter baumanii*	Azo dyes effluents
31.	*Bacillus firmus, Bacillus macerans, Staphylococcus aureus and Klebsiella oxytoca*	Vat dyes, Textile effluents
Heavy Metals Bioremediation		
32.	*Saccharomyces cerevisiae*	Heavy metals, lead, mercury and nickel
33.	*Cunninghamella elegans*	Heavy metals
34.	*Pseudomonas fluorescens* and *Pseudomonas aeruginosa*	Fe^{2+}, Zn^{2+}, Pb^{2+}, Mn^{2+} and Cu^2
35.	*Lysinibacillus sphaericus* CBAM5	Cobalt, copper, chromium and lead
36.	*Microbacterium profundi* strain Shh49T	Fe
37.	*Aspergillus versicolor, A. fumigatus, Paecilomyces* sp., *Paecilomyces* sp., *Terichoderma* sp., *Microsporum* sp., *Cladosporium* sp.	Cadmium
38.	*Geobacter* spp.	Fe (III), U (VI)
39.	*Bacillus safensis* (JX126862) strain (PB-5 and RSA-4)	Cadmium
40.	*Pseudomonas aeruginosa, Aeromonas* sp.	U, Cu, Ni, Cr
41.	*Aerococcus* sp., *Rhodopseudomonas palustris*	Pb, Cr, Cd
Potential biological agents for pesticides.		
42.	*Bacillus, Staphylococcus*	Endosulfan
43.	*Enterobacter*	Chlorpyrifos
44.	*Pseudomonas putida, Acinetobacter* sp., *Arthrobacter sp.*	Ridomil MZ 68 MG, Fitoraz WP 76, Decis 2.5 EC, Malathion
45.	*Acenetobactor sp., Pseudomonas sp., Enterobacter sp. and Photobacterium sp.*	Chlorpyrifos and methyl parathion

Microbial Biotransformation

Biotransformation is an organism's chemical modification(s) on a chemical compound. If the end products of this modification are mineral compounds like CO_2, NH_4^+, or H_2O, the biotransformation is then called mineralization. In other words, biotransformation refers to the chemical alteration of chemical compounds such as nutrients, amino acids, toxins, and drugs in the body of an organism. It also renders non-polar compounds polar, so they are not reabsorbed in renal tubules and excreted [29]. It is seen as an efficient, cost-effective, and easily applicable approach for developing agricultural wastes with the potential to enhance existing bioactive components and synthesize new compounds [30].

Various Processes of Biotransformation

Oxidation

It includes oxidation of alcohols and aldehydes, hydroxylation, epoxidation and dehydrogenation of C-C bonds, oxidative degradation of alkyl, carboxy-alkyl or keto-alkyl chains, oxidative removal of substituents, oxidative deamination, oxidation of hetero-functions and oxidative ring fission [31].

Reductions

Reductions include reducing organic acids, aldehydes, ketones, hetero-functions, dehydroxylation, reductive elimination of substituents, and hydrogenation of C-C bonds.

Hydrolysis

It includes hydrolysis of esters, amines, amides, lactones, ethers, lactams, *etc.*

(iv). Condensation

It includes dehydration, O- and *N*-acylation, glycosidation, esterification, lactomization, and amination.

Isomerization

It includes racemization, rearrangements, migration of double bonds or oxygen functions, and formation of C-C or hetero-atom bonds.

Mixed Reactions

These include hydroxylation with reduction, hydroxylation with oxidation, hydroxylation with side-chain degradation, and rupture of C-C linkages with the oxidation of side chain.

Role of Microbes in Biotransformation

Biotransformation of different pollutants using microorganisms to clean up the polluted environment have achieved so much success in recent years. The catabolic diversity of microbes has played a vital role in the bioremediation of a wide variety of compounds, including pharmaceutical substances, polyaromatic hydrocarbons (PAHs), and hydrocarbons (*e.g.*, oil) and polychlorinated biphenyls (PCBs). The main disadvantage is the toxicity threat these pollutants pose to public health.

Significant developments in molecular techniques, bioinformatics, genomics, and metagenomics are used to amplify the transformation of xenobiotic compounds, and microbes, where primary bacteria with detoxifying properties, have an important role to play in this phase. Vast metabolic diversity, higher growth rate, and high horizontal gene transfer enable them to evolve and adapt to changing environmental conditions. The biotransformation of many xenobiotic chemical agents has been found in aerobic and anaerobic bacterial genera. *Bacillus*, *Pseudomonas*, *Escherichia*, *Rhodococcus*, *Gordonia*, *Moraxella*, *Micrococcus* are members of the aerobic genera, while the anaerobic types include *Methanospirillum*, *Pelatomaculum*, *Syntrophobacter*, *Desulfotomaculum*, *Syntrophus*, *Desulfovibrio* and *Methanosaeta*. *Mycobacterium vaccae* have demonstrated the capability to catabolize acetone, cyclohexane, styrene, benzene, ethylbenzene, propylbenzene, dioxane, and 1,2-dichloroethylene. *Pseudomonas* and *Bacillus* degrade PCB (Polychlorinated Biphenyls) very efficiently [30]. Some strains of *Pseudomonas*, *Acetobacter,* and *Klebsiella* have also been able to bio-fix carcinogenic azo compounds. Anaerobic methanogens (*Methanospirillum hungatei, Methanosaeta concilii, Syntrophobacterfumaroxidens*) are mainly involved in the degradation of Phthalate compounds. Recently *Cunninghamella elegans, Pseudomonas knackmussii,* and *P. pseudoalcaligenes* KF-707 have shown the ability to biotransform potential pollutants bearing the penta fluorosulfanyl (SF$_5^-$) functional group [31].

CONCLUSION

For maintaining the global carbon cycle and renewing our environment, microorganisms have an essential role to play. The various microbial activities are comprehended in biodegradation, bioremediation, and biotransformation. Substances transformed by microorganisms include a wide range of synthetic compounds and other chemical substances like hydrocarbons and heavy metals, which have toxic ecological effects. However, in most cases, this statement is concerned with the potential degradabilities of microorganisms estimated under ideal growth conditions using selected laboratory cultures. Due to a broad range of factors, competition with microorganisms, lack of supply of necessary substrates, adverse environmental conditions (aeration, moisture, pH, temperature), and poor pollutant bioavailability, the biodegradation rate under natural conditions is low. Therefore, environmental biotechnology aims to address and resolve these problems in such a way as to enable the use of microorganisms in bioremediation technologies. For this purpose, it is important to support the activities of indigenous microorganisms in contaminated biotopes and enhance their degrading capabilities through bio-augmentation or biostimulation. Genetic engineering is also used to enhance the ability of microorganisms to biodegrade. However, there are many risks associated with using GEM in the field. Whether or not such methods are eventually effective in the bioremediation of pollutants will make a difference in our ability to minimize waste, eradicate industrial emissions and enjoy a more sustainable future.

CONSENT FOR PUBLICATION

Not applicable.

CONFLICT OF INTEREST

The author declares no conflict of interest, financial or otherwise.

ACKNOWLEDGEMENTS

Encouragement, constant guidance, and moral support of Dr. Khursheed Hussain, Assistant Professor, Potato Tissue Culture Laboratory, Division of Vegetable Science, SKUAST-Kashmir, are duly acknowledged. I also thank Dr. Khalid Z. Masoodi, Assistant Professor, Transcriptomics Laboratory, Division of Plant Biotechnology, SKUAST-Kashmir, who inspires and supports me.

REFERENCES

[1]   Abdulsalam S, Adefila SS, Bugaje IM, Ibrahim S. Bioremediation of soil contaminated with used motor oil in a closed system. Bioremediation and Biodegradation. 2013; pp. 100-72. goo.gl/dBqZRu

[2]   Singh A, Kumar V. Assessment of bioremediation of oil and phenol contents in refinery waste water *via* bacterial consortium. J Pet Environ Biotechnol 2013; 4(3): 1-4.
[http://dx.doi.org/10.4172/2157-7463.1000145]

[3]   Adams GO, Fufeyin PT, Okoro SE, Ehinomen I. Remediation and characterization innovative technologies: Information snapshots: technologies by type. 2015.goo.gl/9XY7ni

[4]   Alexander M. Biodegradation and Bioremediation. San Diego CA. Academic Press 1994.

[5]   Asira Enim Enim. Factors that Determine Bioremediation of Organic Compounds in the Soil 2013; 141-57.goo.gl/aZ2ntP
[http://dx.doi.org/10.5901/ajis.2013.v2n13p125]

[6]   Bennet JW, Wunch KG, Faison BD. Use of fungi biodegradation Manual of environmental microbiology. 2nd ed. Washington, D.C.: ASM Press 2002; pp. 960-71.

[7]   Marinescu M, Dumitru M, Lacatusu A. Biodegradation of Petroleum Hydrocarbons in an Artificial Polluted Soil. Research Journal of Agricultural Science 2009; 41: 2.

[8]   Montagnolli RN, Matos Lopes PR, Bidoia ED. Assessing Bacillus subtilisbiosurfactant effects on the biodegradation of petroleum products. Environ Monit Assess 2015; 1-17.goo.gl/77u5La

[9]   Boopathy R. Factors limiting bioremediation technologies. Bioresour Technol 2000; 74(1): 63-7.goo.gl/eQhPh7
[http://dx.doi.org/10.1016/S0960-8524(99)00144-3]

[10]   Tahri Joutey Nezha, Wifak Bahafid, Hanane Sayel, Naïma El Ghachtouli. Biodegradation: involved microorganisms and genetically engineered microorganisms, biodegradation - life of science, rolando chamy and francisca rosenkranz. TechOpen 2013.
[http://dx.doi.org/10.5772/56194]

[11]   Tekere M. Microbial bioremediation and different bioreactors designs applied, biotechnology and bioengineering. Eduardo Jacob-Lopes and Leila Queiroz Zepka 2019.
[http://dx.doi.org/10.5772/intechopen.83661]

[12]   Phulia V, Jamwal A, Saxena N, Chadha NK. Technologies in aquatic bioremediation. 2013; 65-91.

[13]   Shilpi S. Bioremediation: Features, Strategies and applications. Asian Journal of Pharmacy and Life Science 2012; 202-13. goo.gl/q1JpAh

[14]   Demnerova K, Mackova M, Spevakova V, Beranova K, Kochankova L, *et al.* Two approaches to biological decontamination of groundwater and soil polluted by aromatics characterization of microbial populations. 2005. goo.gl/1ahGcu

[15]   El Fantroussi S, Agathos SN. Is bioaugmentation a feasible strategy for pollutant removal and site remediation? 2005. goo.gl/y6kLsc
[http://dx.doi.org/10.1016/j.mib.2005.04.011]

[16]   Sani I, Safiyanu I, Rita SM. Review on Bioremediation of Oil Spills Using Microbial Approach. IJESR 2015; pp. 41-5.

[17]   Strong PJ, Burgess JE. Treatment methods for wine-related ad distillery wastewaters: a review. Bioremediation Journal 2008; 12: 70-87. goo.gl/HCqJd6

[18]   Tang CY, Fu QS, Criddle CS, Leckie JO. Effect of flux (transmembrane pressure) and membrane properties on fouling and rejection of reverse osmosis and nanofiltration membranes treating perfluorooctane sulfonate containing wastewater. Environ Sci Technol 2007; 41(6): 2008-14.
[http://dx.doi.org/10.1021/es062052f] [PMID: 17410798]

[19]   Macaulay BM. Understanding the behavior of oil-degrading microorganisms to enhance the microbial remediation of spilled petroleum. Appl Ecol Environ Res 2014; 247-62. [Link. https.//goo.gl/JfFVWd].

[20]   Yang SZ, Jin HJ, Wei Z, He RX, Ji YJ, *et al.* Bioremediation of oil spills in cold environments: A review. Pedosphere 2009. goo.gl/S0MQ3
[http://dx.doi.org/10.1016/S1002-0160(09)60128-4]

[21]   Safiyanu I, Isah AA, Abubakar US, Rita Singh M. Review on comparative study on bioremediation for oil spills using microbes. Research Journal of Pharmaceutical, Biological and Chemical Sciences 2015; 783-90. goo.gl/LSj5vR

[22]   Sarang B, Richa K, Ram C. Comparative Study of Bioremediation of Hydrocarbon Fuel. International Journal of Biotechnology and Bioengineering Research 2013; 677-86. goo.gl/CCnV57

[23]   Madhavi GN, Mohini DD. Review paper on – Parameters affecting bioremediation. International journal of life science and pharma research 2012; 77-80. goo.gl/tBP2C6

[24]   Thavasi R, Jayalakshmi S, Banat IM. Application of biosurfactant produced from peanut oil cake by Lactobacillus delbrueckiiin biodegradation of crude oil 2011; 3366-72. goo.gl/FWUrD3

[25]   Jain PK, Bajpai V. Biotechnology of bioremediation- a review. Int J Environ Sci 2012; 535-49. goo.gl/EhLwbz

[26]   Pedro P, Francisco JE, Joao F, Ana L. DNA damage induced by hydroquinone can be prevented by fungal detoxification. 2014.2014.goo.gl/N4Bj8A

[27]   Wu YH, Zhou P, Cheng H, Wang CS, Wu M. Draft genome sequence of Microbacterium profundiShh49T, an Actinobacterium isolated from deep-sea sediment of a polymetallic nodule environment. Genome Announcments 2015; 1-12.1-2.goo.gl/EQ19PW

[28]   Wang Q, Zhang S, Li Y, Klassen W. Potential approaches to improving biodegradation of hydrocarbons for bioremediation of crude oil pollution. 2011.goo.gl/pyGZ6a [http://dx.doi.org/10.4236/jep.2011.21005]

[29]   Kumar A, Bisht BS, Joshi VD, Dhewa T. Review on bioremediation of polluted environment: a management tool. International Journal of Environmental Sciences 2011; 1079-93. goo.gl/P6Xeqc

[30]   Huttel W, Hoffmeister D. Fungal biotransformations in pharmaceutical sciences. In: Hofrichter M, Ed. Industrial applications, the mycota X, 2. Germany: Springer 2010; pp. 293-317.

[31]   Kavanagh E, Winn M, Gabhann CN, O'Connor NK, Beier P, Murphy CD. Microbial biotransformation of aryl sulfanylpentafluorides. Environ Sci Pollut Res Int 2014; 21(1): 753-8. [http://dx.doi.org/10.1007/s11356-013-1985-2] [PMID: 23872898]

[32]   Strong PJ, Burgess JE. Treatment methods for wine-related ad distillery wastewaters: a review. Bioremediation Journal. 2008; 12: pp. 70-87. goo.gl/HCqJd6

[33]   Tang CY, Fu QS, Criddle CS, Leckie JO. Effect of flux (transmembrane pressure) and membrane properties on fouling and rejection of reverse osmosis and nanofiltration membranes treating perfluorooctane sulfonate containing wastewater. Environ Sci Technol 2007; 41(6): 2008-14. [http://dx.doi.org/10.1021/es062052f] [PMID: 17410798]

[34]   Thavasi R, Jayalakshmi S, Banat IM. Application of biosurfactant produced from peanut oil cake by Lactobacillus delbrueckiiin biodegradation of crude oil. Bioresour Technol. 2011; pp. 3366-72. goo.gl/FWUrD3

[35]   Wang Q, Zhang S, Li Y, Klassen W. Potential Approaches to Improving Biodegradation of Hydrocarbons for Bioremediation of Crude Oil Pollution. Environ Protection 2001; 47-55.

[36]   Wu YH, Zhou P, Cheng H, Wang CS, Wu M. Draft genome sequence of Microbacterium profundiShh49T, an Actinobacterium isolated from deep-sea sediment of a polymetallic nodule environment. Genome Announcments. 2015; pp. 1-2. goo.gl/EQ19PW

[37]   Yang SZ, Jin HJ, Wei Z, He RX, Ji YJ, *et al.* Bioremediation of oil spills in cold environments: A review. Pedosphere. 2009. goo.gl/S0MQ3

CHAPTER 8

Bioremediation of Hazardous Organics in Industrial Refuse

Riham Fouzi Zahalan[1,*], **Muhammad Manhal Awad Al-Zoubi**[1] and **Abdel Rahman Mohammad Said Al Tawaha**[2]

[1] *General Commission for Scientific Agricultural Research, Administration of Natural Resource Research, Syria*

[2] *Department of Biological Sciences, Al-Hussein bin Talal University, Maan, Jordan*

Abstract: Increased population and industrial revolution, alongside the wrong agricultural management systems, are putting massive pressure on the natural resources available for human beings. Several international organizations are raising flags and knocking the future risks and costs of exhausting the available natural resources. Soil is categorized as a slowly renewable resource to a limit that made soil experts classify soil as a nonrenewable natural resource. Therefore, soil pollution is among the most important issues discussed at the global level. However, soil remediation is very high costly, time-taking, and needs experts for handling. Bioremediation is considered one of the most promising methods of soil rehabilitation by simulating the behaviour of nature in curing it. With lower costs, noticeable results, and eco-friendly alternative solutions, bioremediation might be the most suitable strategy for polluted lands.

Keywords: Bioremediation, Industrial Revolution, Natural Resources, Soil pollution.

INTRODUCTION

Bioremediation is the term used to describe the processes of biological degradation and deactivation of several toxins and pollutants [1], by using a variety of microorganisms such as fungi and bacteria, *etc.*, or by using some specific sorts of plants [2]. Bioremediation is among the most promising mechanisms, representing [3] about 32% of the allowable technologies adopted in contaminated soils to face the hazards of these pollutants (Fig. **1**). Organic toxins are a dangerous type of contaminants that usually reach the agricultural lands loaded with agrochemicals, such as pesticides, and synthetic fertilizers, in addition to the use of organic materials as soil amendments, which are derived initially

---

[*] **Corresponding author Riham F. Zahalan:** General Commission for Scientific Agricultural Research, Administration of Natural Resource Research, Damascus, Syria; E-mail: rihamzah7@gmail.com

from wastes or organic materials in touch with pollution sources like synthetic dyes, wood preservatives, *etc.*

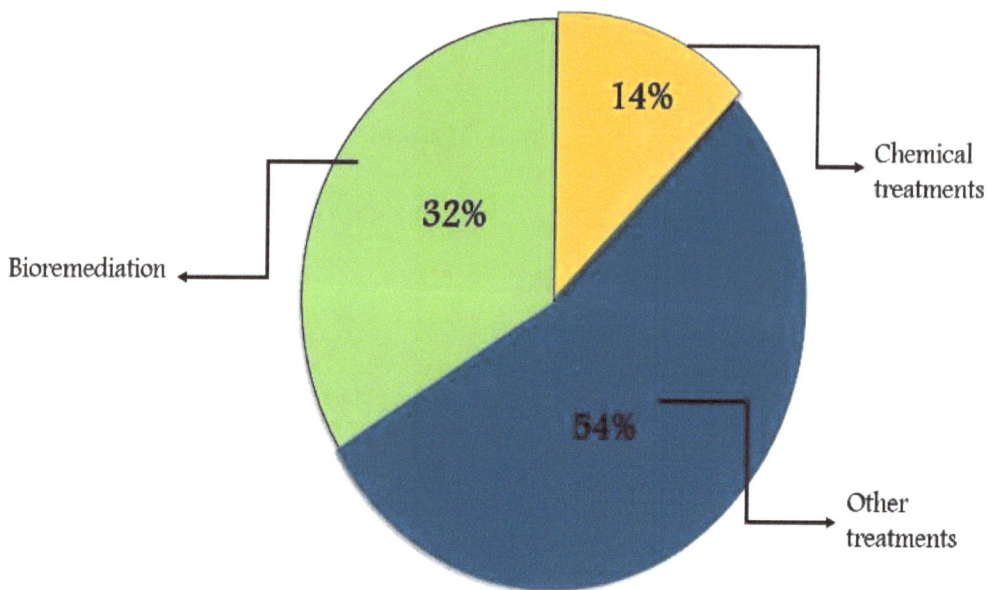

Fig. (1). Percentages of different types of soil remediation treatments [3].

These types of contaminants are a concern for soil and environmental experts due to their harmful effects on the soil-plant-animal-human chain as carcinogenic toxins and teratogenic factors [4].

If we explore the recent trends of dealing with and using the accumulated and increased amounts of waste, we could easily notice the shift in understanding and accepting the use of these wastes in agricultural lands. This use that began several decades ago is increasing dramatically and is produced as an Eco-friendly and alternative solution, to the increased accumulated wastes, as a natural result of the increased population and industrial revolution in the modern world. However, the results of several research works in different soil types and parts in the world proved the positive effect of this use, yet it should be monitored and wisely controlled. Therefore, it would not be a new source of contamination in arable lands.

Principles and Importance of Bioremediation

The principle of bioremediation is based on transforming the pollutant compounds by living organisms through the natural metabolic processes in their bodies [5]. When microorganisms are moved to the polluted soil, they start to secrete specific kinds of extracellular enzymes, which in turn will attack the existing contaminants and then convert them into unavailable forms for plant absorption or harmless substances to the environment and humans at the same time. The main idea behind adapting these kind of remediation processe and techniques is due to the increased awareness of pollution risks and the raised flags to protect natural resources. Bioremediation is trying to simulate nature in curing itself. However, the importance of bioremediation goes beyond the safety and natural use to the economic effects and costs.

In recent years, there have been increased demands to protect the soil biodiversity, increasing global awareness of their essential role in the human food chain and in saving the natural balance. The use of bioremediation goes along with that aim since it initially depends on increasing the soil content of microorganisms and rebuilding the numbers and the types of these microorganisms in degraded and polluted soils.

Types of Bioremediation

Bioremediation techniques include two main branches: *ex-situ* and *in-situ*, indicating where the treatment process will take place and happen (Fig. **2**).

Ex-situ Bioremediation Techniques

Ex-situ technique: This type of bioremediation depends on removing the pollutants from polluting sites by applying specific treatments out of the original site [6].

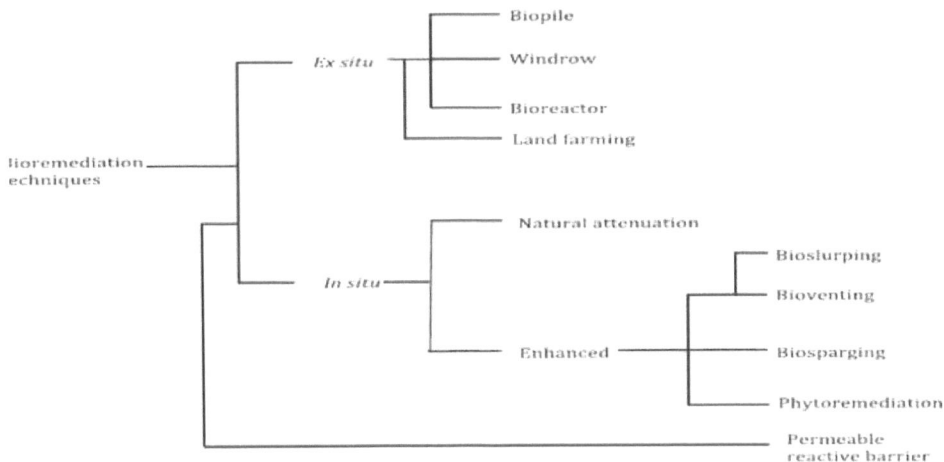

Fig. (2). Types of Bioremediation [6].

This transporting process is highly dependent on the geographical and geological characteristics of the polluted site, depth, severity, and type of pollutants. *Ex-situ* techniques include Bio-pile, Windrow, Bioreactor, and Land farming.

Bio-pile: This type is known for cost-effectiveness and suitability for polluted sites in cold regions [7]. Bio-pile is based on aggregation and accumulation of the excavated polluted soil, followed by treatment of the piled soil above the ground by applying controlled actions [8], such as the use of nutrients, irrigation, and aeration to induce and increase microbial activities. This process will require good bed treatment, irrigation, and leachate collection systems. Among the successful experiments that used the bio-pile method was the trial of a study [9], which used 5 &10% of mature compost with microbial consortia. This experiment achieved a 90.7% reduction in the total petroleum hydrocarbon at conditions of low temperatures [7]. Also, a 71% reduction was reported in the total hydrocarbon after 50 days of the study. The bio-pile method could be scaled up by controlling the temperature of the air injected into the bio-pile design to enhance the bioremediation process. On the other hand, one of the essential advantages of this method is its capacity to treat large amounts of polluted soil in small spaces.

Windrow: This type of bioremediation technique depends on pilling the polluted soil, which will undergo a controlled process of periodic turning to induce the microorganisms' activities and enhance the degradation process.

This method also includes adding water, which with the turning process would ensure the uniformity of pollutants' distribution, the aeration, and the nutrients' distribution in the treated soil. Despite the high efficiency that could be achieved in this method, the possibility of releasing greenhouse gas (CH_4) could be a limiting factor, especially in soils polluted with toxic volatiles [6].

Bioreactor: This type of bioremediation needs a particular container or a vessel, which works as an apparatus in which biological reactions and processes will carry out. Polluted soil (dry or slurry) is fed to the vessel and converted into specific products after a series of biological reactions. The bioreactor method helps control all conditions to support the natural biological process. However, the essential characteristic of this type is its capacity to ensure controlled, effective and maximum pollutant bioavailability and mass transfer (the contact between the pollutants and microbes), which is usually one of the limiting factors faced . Several researchers approved the effectiveness of the bioreactor method for soil and water polluted with volatile organic compounds such as benzene, toluene, ethylbenzene, and xylene in water [10], for 2,4-dichlorophenoxyacetic acid in soil [11], and for carbofuran [12], and 2,4,6-trinitro phenyl methylintramine [13] in polluted soil.

Land Farming: This type is on the line between *ex-situ* and *in-situ* techniques since both cases could be implied. In general, it relies on enhancing the bioavailability of autochthonous soil organisms [14]. A study suggested the depth of the polluted layer to be the factor in which soil would be transported to the ground surface or not (*Ex-situ* or *in-situ*). A simple and costless method that does not need special equipment yet depends on tillage, irrigation, and aeration as significant operations.

Although land farming involves the advantages of several methods related to the implementation in cold areas, the large amounts of treated soil, and the possibility of scaling up, nevertheless, it has several limitations since it is done in ample space, uncontrolled or unfavorable environmental conditions that are less effective for inorganic pollutants' removal and toxic volatiles. All these limitations made it time-consuming and less efficient compared with other methods.

In-situ Bioremediation Techniques

They are used where the treatment operations conducted in the polluted soil are carried out at the pollution site; therefore, there is no need for soil disturbance or excavation process. This technique is suitable for various pollutants such as heavy metals, hydrocarbons, chlorinated solvents, *etc*. [6]. Since these methods are conducted in the field, several environmental conditions would be the main factors that influence their affectivity [15]. It is also important to mention the valuable role of soil structure of the polluted site, particularly the properties related to soil aeration, such as soil porosity.

Bioventing: This technology method tries to affect and control the aeration conditions of the treated sites, which could be achieved by injecting an airflow to the unsaturated zones [6]. This could also involve adding moisture and necessary nutrients to enhance the bioremediation process [15]. Previous research works reported positive results when bioventing, referenced as [16], which reported a percentage exceeding 93% of pollutants removal at the end of the process in soil polluted with phenanthrene [17]. Also, in a study, bioventing for treating soil from light petroleum products was used, while [18] another study used it for sites polluted with toluene. The intense intervals, repeating of air injections, and the air uniformity that could be achieved are the basic parameters to ensure successful bioventing. However, soil permeability could be a limiting factor and also the possibility of releasing gas emissions in soils contaminated with volatile organic compounds. Some researchers tried to overcome these problems by injecting pure oxygen in soils with poor permeability and combining other helping techniques such as bio trickling filters with bioventing to decrease gas emissions [19].

Bioslurping: This type of bioremediation is very similar to the bioventing method but is usually used for removing free products such as volatile and semi-volatile organic compounds and non-aqueous phase liquids by applying oxygen as a microorganism's enhancer to increase bioremediation process. In the bioslurping method, vacuum pumping is required to remove the accumulated pollutants that are separated and gathered on water or air surface. So similar to bioventing, moisture conditions, the medium permeability, and the design of the vacuuming system have a substantial effect on bioslurping success [15].

Biosparging: Like the previous methods, this type depends on air injection to enhance the microorganisms' activity, with one main difference being that the injected air should be in the saturated zones of the treated site. This aims to use the power of the injected air flow to push the volatile organic compounds from saturated zones toward the unsaturated zones, where biodegradation will occur.

In this way, the significant challenges that could limit its effectivity are the correct determination of the most suitable saturated zones to inject, controlling the directions of the injected air, soil permeability conditions, and biodegradability of the existed pollutants [15]. Many researchers [20] used this method and reported positive results in soils polluted with ethylbenzene, benzene, and toluene.

Permeable Reactive Barrier: This method was designed for groundwater pollution. Most researchers classified it as a physical method rather than a biological or chemical one based on the basic treating mechanisms. A specific medium (usually a zero-valent iron) is used as a permanent or semi-permanent reactive barrier, which will be submerged in the path of the polluted groundwater, to trap the pollutants but release the water flow. What determines the effectiveness of this method the most is the materials used to build the barrier, its suitability to the pollutants' type, and other environmental characteristics of the polluted site. One thing that should be considered in this method is ensuring the health effects, especially the barrier material because dealing with groundwater is a very sensitive task. In contrast, other researchers were interested in the economic cost of manufacturing, stability, and the sustainability of the barrier [21].

Intrinsic Bioremediation: The absence of external forces or human interventions distinguishes this method from others. In this type, a biodegradation process is carried out by both aerobic and anaerobic microbes and could suit a wide range of pollutants, even the recalcitrant ones [6]. Although this method is less expensive in terms of costs as well as in terms of human labor, yet still needs to be under control to ensure effectiveness and sustainability of biodegradation processes. To establish an excellent intrinsic process, experts must demonstrate effective and suitable microbes usually isolated from contaminated sites and study their potential to break down the targeted pollutants.

During the bioremediation period, experts also should monitor the contaminants lost from treated soil toward groundwater, for example. Significant limitations related to this method include: It is time-consuming compared with other options; this time differs according to the pollutants concentrations and the target remediation level. It needs excessive study, and several factors should be covered and determined precisely before starting a bioremediation process. Since this part is also time-consuming, experts should consider that the time needed to finish the remediation process should not exceed the predicted time the concentration of pollutants may reach the groundwater or human/animal food.

Phytoremediation: This method depends on soil-plant interaction for pollutants' removal, where several mechanisms could be involved in the process, such as uptake, filtration [22], volatilization, degradation [23] and accumulation [24], and others [6]; according to the type of contaminants and the polluted site characteristics. Therefore additional factors should be considered related mainly to the plant characteristics, including the type of plant, root system, growth rate, plant adaptability and sensitivity to pests and diseases, and the planned final fate of these plants. Plants that usually grow for their wood (inedible final products) and plants unavailable for animal feed are preferred in this method, which should be incinerated or buried in specific ways to ensure an integrated bioremediation process. The advantages of this method include the low cost and the ability to remove a wide range of pollutants (organic wastes, heavy metals, *etc.*). Several studies used and approved this method as an Eco-friendly and effective remediation process for heavy metals [25, 26], Gasoline [27], and polychlorinated fluoranthene [28].

Factors that Affect the Bioremediation Effectiveness

Soil conditions, environmental factors, microorganisms' status, and contaminants (forms and availability) are the three basics of the bioremediation process. Therefore, any factor that can influence one of these fundamentals will significantly affect the overall bioremediation process and the final results.

Microorganism Affecting Conditions: These conditions include factors that affect the microbial effectiveness in the polluted soil. The first to be considered is the microbial population. Achieving satisfactory results requires adding an adequate amount of microbial community, which should also consider the level, type and concentration of available pollutants. When planning a bioremediation process, available contaminants and suitable types of organisms should be among the basics (Organisms with the physiological and metabolic capabilities to degrade the pollutants).

Adequate knowledge is required to ensure positive and effective outcomes. For example, Chlorinated organics are known for resisting the microbial attack; in this case, the use of bioremediation will be useless and a waste of time, money, and effort. On the other hand, certain species of earthworms such as (*Perionyx excavates – Eurilus engeniae* and *Eisenia fetida*) are known for their influential role in extracting and eliminating several heavy metals (cadmium – lead – copper-manganese – zinc) when used in producing composts [29]. Therefore, using them while producing and preparing organic amendments, especially waste, could be a wise choice.

Bioremediation Conditions: This sector usually includes the conditions of the growth environment and many physical-chemical environmental parameters. These conditions directly impact microorganisms' growth and affectivity; therefore should be controlled to ensure optimum suitability. This could be achieved by "bio-stimulation," which includes additions of enhancing factors such as nutrients (Nitrogen- Phosphorous-Carbon) and oxygen [5]. The first to control are the pH and temperature of the growth environment. While most known microorganisms usually prefer to grow in a neutral pH range, there is a different behavior toward growth at medium temperature. Microorganisms differ from mesophilic to thermophilic or thermotropic species [30]. When thermophilic microorganisms were used to report a high degradation rate of organic compounds in wastes and wastewaters (60-70 °C). Moisture and nutrient availability have a vital role in helping Microorganisms grow, build their bodies [31], and then create the necessary enzymes that will break down the pollutants and accelerate biodegrading. Soil type and structure (Porosity, permeability, waterlogging, and other soil conditions) play an essential role in the biodegradation rate, especially the soil porosity and its capacity to hold water. Soil porosity directly affects biodegradation; eventually, it could be a limiting factor in some cases [5]. Chlorinated compounds could be degraded only under anaerobic conditions, while hydrocarbons could be degraded easily under aerobic conditions. Improving and controlling soil conditions could be achieved by adding soil amendments such as organic matter, gypsum, *etc.* [31], or by adding oxygen or hydrogen peroxide, which is seven times more soluble in water than oxygen [30].

Bioavailability of Contaminants: Bioremediation mainly depends on breaking down the pollutants, which differ in their bioavailability to be broken down; therefore, any external enhancing factors or materials could increase and accelerate biodegradation processes. This could be done by adding bio-surfactants or synthetic detergents to the polluted sites [30]. Found an increased degradation rate in sites polluted with polycyclic aromatic hydrocarbons after adding biodegradable solvents, which assist and increase the pollutant's adsorption.

Other limitations include the availability of qualified and trained experts, adequate knowledge about the polluted site, and the characteristics of the microbes and the pollutants (microbiology-chemistry).

On the other hand, collecting the necessary information related to the environmental conditions of growth medium (soil science- geology- hydrology), exhaustive investigations and good planning (engineering- project management) are considered among the basics. There is a strong need for combined efforts and expertise to use bioremediation techniques successfully.

Period of remediation could be added to these limitations. This period could last from several months to several years based on the primary situation of the polluted site and the method of remediation, side by side with the available potential (The cost of the process and the technical requirements).

CONCLUSION

Pollution of natural resources is affecting the quality and sustainability of human life on this earth. With the industrial revolution, new types of pollution and complex pollutants have increased that risk. Therefore, environmental and agricultural experts are trying to find practical solutions. Bioremediation has proved to be a possible remediation solution for different types of pollutants and can improve soil biodiversity at the same time. The bioremediation approach is now a promising technology for effective environmental clean-up.

CONSENT FOR PUBLICATION

Not applicable.

CONFLICT OF INTEREST

The author declares no conflict of interest, financial or otherwise.

ACKNOWLEDGEMENT

Declared none.

REFERENCES

[1]   Evans GM, Furlong JC. Environmental biotechnology: Theory and applications. John Wiley and Sons, Chichester. UK. 2003.

[2]   Nealson KH. Harnessing microbial appetites for remediation. Nat Biotechnol 2003; 21(3): 243-4.
        [http://dx.doi.org/10.1038/nbt0303-243] [PMID: 12610569]

[3]   EPA REACH IT. Remediation and characterization innovative technologies: Information snapshots:
        technologies by type, 2004, www.epareachit.org.

[4]   Brar SK, Verma M, Surampalli RY, *et al.* Bioremediation of hazardous wastes. Pract Period Hazard
        Toxic Radioact Waste Manage 2006; 10(2): 59-72.
        [http://dx.doi.org/10.1061/(ASCE)1090-025X(2006)10:2(59)]

[5]   Gupta C, Prakash D. Novel bioremediation methods in waste management. In: Rathoure, A.K and
        V,K, Dhatwalia, Toxicity and waste management using bioremediation, Advances in Environmental
        Engineering and Green Technologies Book Series, 2016; 141-57.
        [http://dx.doi.org/10.4018/978-1-4666-9734-8.ch007]

[6]   Azubuike CC, Chikere CB, Okpokwasili GC. Bioremediation techniques–classification based on site
        of application: principles, advantages, limitations and prospects. World J Microbiol Biotechnol 2016;
        32(11): 180-98.
        [http://dx.doi.org/10.1007/s11274-016-2137-x] [PMID: 27638318]

[7]   Dias RL, Ruberto L, Calabró A, Balbo AL, Del Panno MT, Mac Cormack WP. Hydrocarbon removal
        and bacterial community structure in on-site biostimulated biopile systems designed for
        bioremediation of diesel-contaminated Antarctic soil. Polar Biol 2015; 38(5): 677-87.
        [http://dx.doi.org/10.1007/s00300-014-1630-7]

[8]   Whelan MJ, Coulon F, Hince G, *et al.* Fate and transport of petroleum hydrocarbons in engineered
        biopiles in polar regions. Chemosphere 2015; 131: 232-40.
        [http://dx.doi.org/10.1016/j.chemosphere.2014.10.088] [PMID: 25563162]

[9]   Gomez F, Sartaj M. Optimization of field scale biopiles for bioremediation of petroleum hydrocarbon
        contaminated soil at low temperature conditions by response surface methodology (RSM). Int
        Biodeterior Biodegradation 2014; 89: 103-9.
        [http://dx.doi.org/10.1016/j.ibiod.2014.01.010]

[10]   Firmino PIM, Farias RS, Barros AN, *et al.* Understanding the anaerobic BTEX removal in continuous-
        flow bioreactors for *ex situ* bioremediation purposes. Chem Eng J 2015; 281: 272-80.
        [http://dx.doi.org/10.1016/j.cej.2015.06.106]

[11]   Mustafa YA, Abdul-Hameed HM, Razak ZA. M and Razak, Z.A, Biodegradation of 2,4-
        dichlorophenoxyacetic acid contaminated soil in roller slurry bioreactor. Clean (Weinh) 2015; 43(8):
        1241-7.
        [http://dx.doi.org/10.1002/clen.201400623]

[12]   Plangklang P, Reungsang A. Bioaugmentation of carbofuran by Burkholderia cepacia PCL3 in a
        bioslurry phase sequencing batch reactor. Process Biochem 2010; 45(2): 230-8.
        [http://dx.doi.org/10.1016/j.procbio.2009.09.013]

[13]   Fuller ME, Kruczek J, Schuster RL, Sheehan PL, Arienti PM. Bioslurry treatment for soils
        contaminated with very high concentrations of 2,4,6-trinitrophenylmethylnitramine (tetryl). J Hazard
        Mater 2003; 100(1-3): 245-57.
        [http://dx.doi.org/10.1016/S0304-3894(03)00115-8] [PMID: 12835026]

[14]   Nikolopoulou YA, Abdul-Hameed H. M and Razak, Z.A, Biodegradation of 2,4-
        dichlorophenoxyacetic acid contaminated soil in a roller slurry bioreactor. Clean (Weinh) 2015; 43:
        1115-266.

[15]   Gupta C, Prakash D. Novel bioremediation methods in waste management. In: Rathoure, A.K and
        V,K, Dhatwalia, Toxicity and waste management using bioremediation, Advances in Environmental
        Engineering and Green Technologies Book Series 2016; 141-57.

[16]   Frutos FJG, Escolano O, García S, Babín M, Fernández MD. Bioventing remediation and ecotoxicity
        evaluation of phenanthrene-contaminated soil. J Hazard Mater 2010; 183(1-3): 806-13.

[http://dx.doi.org/10.1016/j.jhazmat.2010.07.098] [PMID: 20800967]

[17]   Höhener P, Ponsin V. *In situ* vadose zone bioremediation. Curr Opin Biotechnol 2014; 27: 1-7.
[http://dx.doi.org/10.1016/j.copbio.2013.08.018] [PMID: 24863890]

[18]   Sui H, Li X. Modeling for volatilization and bioremediation of toluene-contaminated soil by bioventing. Chin J Chem Eng 2011; 19(2): 340-8.
[http://dx.doi.org/10.1016/S1004-9541(11)60174-2]

[19]   Magalhaes SMC, Jorge RMF, Castro PML. Magalhaes, S.M.C, Jorge, R.M.F and Castro, P.M.L, Investigations into the application of a combination of bioventing and biotrickling filter technologies for soil decontamination processes: a transition regime between bioventing and soil vapour extraction, Journal of Hazard Mater., 2009; 170: 711-5.

[20]   Kao CM, Chen CY, Chen SC, Chien HY, Chen YL. Application of *in situ* biosparging to remediate a petroleum-hydrocarbon spill site: Field and microbial evaluation. Chemosphere 2008; 70(8): 1492-9.
[http://dx.doi.org/10.1016/j.chemosphere.2007.08.029] [PMID: 17950413]

[21]   Obiri-Nyarko F, Grajales-Mesa SJ, Malina G. An overview of permeable reactive barriers for *in situ* sustainable groundwater remediation. Chemosphere 2014; 111: 243-59.
[http://dx.doi.org/10.1016/j.chemosphere.2014.03.112] [PMID: 24997925]

[22]   Yadav BK, Siebel MA, van Bruggen JJA. Rhizofiltration of a heavy metal (Lead) containing wastewater using the wetland plant *Carex pendula.* Clean (Weinh) 2011; 39(5): 467-74.
[http://dx.doi.org/10.1002/clen.201000385]

[23]   Di Gregorio S, Gentini A, Siracusa G, Becarelli S, Azaizeh H, Lorenzi R. Phytomediated biostimulation of the autochthonous bacterial community for the acceleration of the depletion of polycyclic aromatic hydrocarbons in contaminated sediments. BioMed Res Int 2014; 2014: 1-11.
[http://dx.doi.org/10.1155/2014/891630] [PMID: 25170516]

[24]   Wang J, Koo Y, Alexander A, *et al.* Phytostimulation of poplars and Arabidopsis exposed to silver nanoparticles and Ag^+ at sublethal concentrations. Environ Sci Technol 2013; 47(10): 5442-9.
[http://dx.doi.org/10.1021/es4004334] [PMID: 23631766]

[25]   Mesa J, Rodríguez-Llorente ID, Pajuelo E, *et al.* Moving closer towards restoration of contaminated estuaries: Bioaugmentation with autochthonous rhizobacteria improves metal rhizoaccumulation in native Spartina maritima. J Hazard Mater 2015; 300: 263-71.
[http://dx.doi.org/10.1016/j.jhazmat.2015.07.006] [PMID: 26188869]

[26]   Elias SH, Mohamed M, Nor-Anuar A, *et al.* Ceramic industry wastewater treatment by rhizofiltration system application of water hyacinth bioremediation. Inst Integ Omics Appl Biotechnol Journal 2014; 5: 6-14.

[27]   Fadhile Almansoory A, Abu Hasan H, Idris M, Sheikh Abdullah SR, Anuar N. Potential application of a biosurfactant in phytoremediation technology for treatment of gasoline-contaminated soil. Ecol Eng 2015; 84: 113-20.
[http://dx.doi.org/10.1016/j.ecoleng.2015.08.001]

[28]   Somtrakoon K, Chouychai W, Lee H. Phytoremediation of anthracene and fluoranthene contaminated soil by *Luffa acutangula.* Maejo Int J Sci 2014; 8: 221-31.

[29]   Kamarudheen N, George C, Bose I, Sathiavelu M, Sathiavelu A. Earthworm mediated bioremediation of phenol. Asian Journal of Microbiology &Environmental Sciences 2014; 16(2): 449-53.

[30]   Sharma J. Advantages and limitations of *in situ* methods of bioremediation. Recent Advances in Biology and Medicine 2019; 5: 1-9.
[http://dx.doi.org/10.18639/RABM.2019.955923]

[31]   Prescott LM, Harley JP, Klein DA. Microbiology, Fundamentals of applied. Microbiology 2002; 2: 1012-4.

CHAPTER 9

Role of Microbial Biofilms in Bioremediation

Pratibha Vyas[1,*], **Amrita Kumari Rana**[1] and **Kunwarpreet Kaur**[1]

[1] *Department of Microbiology, College of Basic Sciences and Humanities, Punjab Agricultural University, Ludhiana-141004, Punjab, India*

Abstract: Various types of toxic chemicals and waste materials generated from different industrial processes have created environmental pollution leading to a challenge for healthy human life globally. There is a need to develop strategies for environmental renewal and maintaining healthy life. Bioremediation has emerged as a promising and eco-friendly approach as microorganisms have vast potential to remove toxic pollutants from the environment. Microbial biofilms can be used successfully for removing environmental pollutants because of their ability to degrade, absorb and immobilize a large number of pollutants from various sources. During bioremediation, metabolic activities of biofilm-forming microorganisms are used for degrading toxic environmental pollutants. Though information on the use of microbial biofilms for bioremediation is limited, biofilms have proved to be highly effective in bioremediation. The present chapter focuses on the application and potential of microbial biofilms for the removal of environmental pollutants for sustainable development.

Keywords: Biofilms, Bioremediation, Dioxins, Environment, Heavy metals, Pollutants, Pharmaceuticals, Personal Care Products, Polycyclic Aromatic Hydrocarbons, Polychlorinated Biphenyls.

INTRODUCTION

During the last few decades, a large amount of wastes generated from various industries have led to the problems of contaminated sediments and their disposal in aquatic environments worldwide. In recent years, extensive efforts have been made to manage these pollutants. However, the employed techniques, including physicochemical methods, are expensive and aggravate the problem due to the conversion of wastes into persistent forms or more toxic metabolites. Therefore, cost-effective and eco-friendly alternatives are required to remove these pollutants. Conversion of these toxic pollutants into less or non-toxic compounds with microbial interventions is a possible solution, whereby microorganisms use

---

* **Corresponding author Pratibha Vyas:** Department of Microbiology, College of Basic Sciences and Humanities, Punjab Agricultural University, Ludhiana-141004, Punjab, India; E-mail: vyasp2000@pau.edu

Arun Karnwal & Abdel Rahman Mohammad Said Al-Tawaha (Eds.)

these wastes as carbon, nitrogen, and energy sources. The ability of microorganisms to degrade/neutralize a variety of inorganic and organic pollutants has been used for many years. The free-living microbial populations, though, have the capability of degrading pollutants; however, their low activity due to the nutrient limitations and unfavorable environment can slow down the process [1]. Structured microbial communities known as biofilms overcome these problems as the microbial cells in the biofilm are embedded in the extracellular polymeric matrix [2, 3]. Biofilms have a vast potential for remediating toxic pollutants because of their ability to grow in high numbers and immobilize pollutants.

Additionally, biofilms as a bioremediation technique are advantageous since a biofilm can contain various organisms with different metabolic properties, thus simultaneously metabolizing different pollutants [4]. In recent years, research has been carried out to treat different wastes and pollutants of various habitats with biofilms, thereby suggesting the potential of biofilm communities in the bioremediation of pollutants [5 - 8]. In the natural environment, the native biofilm communities continuously carry out the process of pollutant detoxification as a part of nutrient cycling; however, the process can be enhanced by facilitating biofilm formation [9].

The present chapter aims to provide an overview of the strategies used for the remediation of various pollutants, including persistent organic pollutants, heavy metals, and pharmaceuticals using biofilms. In addition, the limitations of bioremediation using biofilms are also discussed.

Major Pollutants

Many different organic pollutants (*i.e.,* pesticides, surfactants, pharmaceuticals, biocides, polychlorinated biphenyls, polychlorinated dibenzofurans, and polychlorinated dibenzo-p-dioxins) contaminate the environment intentionally or unintentionally due to various industrial and human activities. The majorly found organic pollutants in soil are alkenes, cycloalkenes, and alkanes in the category of oil hydrocarbons [10]; endrin, heptachlor, captan, benomyl, and endosulfan as insecticides [11]; acetochlor, atrazine, alachlor, and bifenoxas pesticides and herbicides [12]; penconazole, procymidone, metalaxyl and lindane as fungicides [13] and xylene, toluene, benzene and ethylbenzene in the category monomeric aromatic hydrocarbons [14]. Many of these soil pollutants are toxic, bioaccumulative, and persistent [15]. Moreover, pharmaceuticals and personal care products enter groundwater through leaching from the soil. The residues of many pesticides remain active in the environment over the years, causing severe health hazards in non-targeted animals due to their solubility in fats and bioaccumulation. The residues of pesticides also create a nutrient imbalance and reduce the quality of agricultural produce. In addition, the utilization of these pesticide residues has been shown to cause cancer, mutations, morphological changes in cells, and other health hazards in animals and human beings.

Furthermore, chemical fertilizers have also led to groundwater contamination with nitrates. The leaching of chemical fertilizers from soil to various water systems also results in eutrophication due to excessive algae growth. These agrochemicals pose a serious environmental threat because of their toxicity and accidental spills, leaching, and leakage of the stored bulk systems. Various types of metals from industrial wastes constitute inorganic pollutants in the environment. The primary contaminants arising from industrial activities are sulphur dioxide and arsenic fluids [16]. Additionally, heavy metals including arsenic, copper, mercury, lead, nickel, and cadmium from industrial wastewater enter into soil and affect the soil's productivity and also cause microorganisms' toxicity.

Bioremediation

Bioremediation, an ingenious and promising technique, is relevant for removing and depleting pollutants from water, air, and soil. Bioremediation technology utilizes microorganisms able to degrade or treat various pollutants, including xenobiotics, pesticides, herbicides, volatile organic compounds, aromatic hydrocarbons, heavy metals, petroleum products, radionuclides, *etc.* [17]. Some microbial species useful in bioremediation are *Pseudomonas*, *Bacillus*, *Arthrobacter*, *Corynebacterium*, *Methosinus*, *Rhodococcus*, *Mycobacterium*, *Nocardia*, *Azotobacter*, *Alcaligenes*, *etc.* Because of the ability of microorganisms to adapt to various environmental conditions due to diverse catabolic pathways, they are highly efficient in removing pollutants [18]. The microbes used in the bioremediation process can tolerate various extremes of pH, temperatures, heavy metals, and radioactive compounds. Bioremediation is preferred over various generally practiced techniques like soil washing in view of its sustainable nature as other procedures are observed to kill or harm habitats and soil organic matter, jeopardize soil health in the long run and deteriorate post-remediation productivity of soil [19]. Sustainability benefits due to the phenomenon of bioremediation include a notable decrease in cost, an increase in the safety of workers, a comparatively smaller life cycle than traditional remediation methods, environmental footprints, and maximum social, economic, and environmental benefits of soil remediation [20]. Agricultural soils tend to accommodate a higher concentration of microorganisms which are able to survive in contaminated soil environments due to the presence of specific genes [21 - 23]. Considering the application's site, suitable remediation technologies are grouped as *ex-situ* or *in-situ.* The criterion for selection of the bioremediation technique includes the type of environment, the nature of the pollutant, pollution level, cost, and environmental policies [24, 25]. Performance criteria like concentrations of nutrients and oxygen, temperature, pH, and various abiotic factors are also given significant considerations along with the selection criteria prior to a remediation project which in turn determine the accomplishment of bioremediation processes [26]. Excavating contaminants from the sites and subsequently transferring them to some other place for treatment are the main steps involved in the technique of *ex-situ* bioremediation. The treatment cost, depth of pollution, pollutant type, the level of pollution, geography, and geology of the contaminated site are usually considered when choosing an appropriate *ex-situ* bioremediation technique, along with some performance criteria. Some of the techniques of *ex-situ* bioremediation are Biopile, Windrows, Land farming, Bioreactors, *etc.* Pollutant inhomogeneity due to excavating processes involved in remediation can be restrained by successfully optimizing specific parameters (temperature, pH, mixing). This technology does not require substantial evaluation of contaminated sites earlier than bioremediation; making the preliminary phase shorter, less strenuous and less costly which are considered its unique advantages.

Treatment of polluted substances at the site of pollution involves the use of the technique for *in situ* bioremediation. There is either significantly less or no disturbance to the structure of the soil and this technology is less costly than *ex-situ* techniques as any excavation does not accompany it. Nevertheless, the improvement of microbial activities involves on-site setting up of advanced equipment during bioremediation, and the cost of design might turn out to be a major concern. Specific techniques used for *in situ* bioremediation like bioventing, biosparging, bioslurping, permeable reactive barrier, and phytoremediation might need some enhancement, while some techniques like intrinsic bioremediation or natural attenuation do not require any improvement [27]. Notably, the most critical environmental conditions required for a successful *in situ* bioremediation include the availability of nutrients, electron acceptor status, pH, moisture content, and temperature. Soil porosity firmly impacts the implementation of *in situ* bioremediation to any contaminated site, unlike *ex situ* bioremediation techniques.

Natural attenuation or intrinsic bioremediation is another *in situ* bioremediation technique involving passive remediation of contaminated areas without any human intervention. Both aerobic and anaerobic microbial processes are involved in the biodegradation of polluting substances, including those that are recalcitrant. Compared to other *in situ* techniques, this technique is less costly due to the absence of external force. It is likely to take an extended period to lower the concentration of contaminants which is one of the major limitations of this process, as there is no incorporation of any external force to hasten the remediation process. Moreover, the intrinsic bioremediation cannot remove polyaromatic hydrocarbons adequately and thereby soil eco-toxicity is not reduced to a greater extent [28].

Since bioremediation is dependent on microbial processes, biostimulation and bioaugmentation are two required methods to expedite the activity of microorganisms in contaminated sites. The stimulation of activities of autochthonous microbes by adding nutrients or substrates to a contaminated sample is called biostimulation. As microorganisms are all-pervasive, the number and metabolic activity of pollutant degraders might vary in reaction to the concentration of contaminants; so, the agro-industrial wastes can be used with suitable nutrient composition to resolve the challenge of limitation of nutrients in most contaminated sites. Degradation of pollutants with microbial consortium utilization is more evident and efficient than single culture [29]. When isolates are mixed, they will bring about harmonious effects due to the absolute and swift degradation of contaminants by their metabolic diversities, which might originate from their isolation sources, adaptation progression, or composition of pollutants [30].

Appropriate prediction of the distribution of contaminants and the identification of pollution sources are crucial to expedite global soil mapping and establish regional models. Field-wide assessments should rather be employed as with the increase in plot size; the variability decreases, due to which the observed efficacy of bioremediation has been a bit unpredictable [31]. The field stations must understand the mechanisms rendering heavy-metal-contaminated sites challenging for treatment. In order to widen the applications of bioremediation technologies to include more sites and decrease clean-up time, more intense research is required. Detailed studies are also required to develop a better technology to get a cleaner environment and safer future for our next generations [32].

Biofilms

Microbial cells attached to an inert or living surface forming well-structured communities enclosed in an extracellular polysaccharide matrix (EPM, a self-produced polymeric matrix) are called biofilms. Biofilms trap nutrients required for microbial growth and help in the detachment of the microbial cells in the aquatic environment. The first step in forming a biofilm is the attachment of microbial cells with a surface (Fig. **1**). Further, biofilm-specific genes are expressed, leading to the synthesis of proteins acting as signaling molecules, thereby initiating matrix formation.

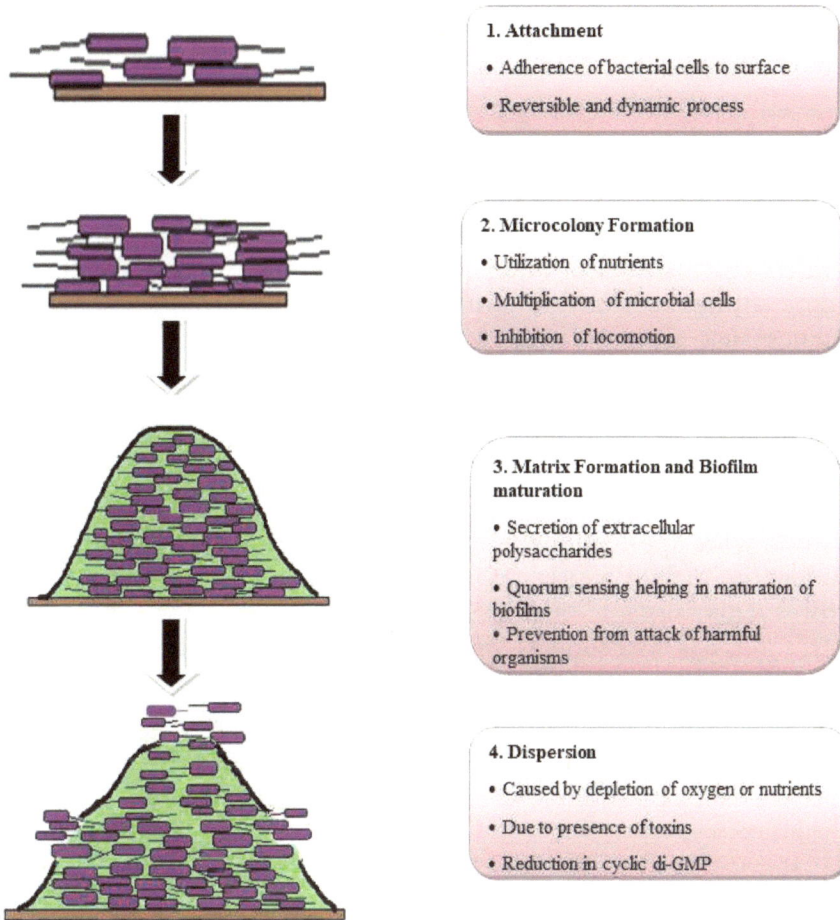

1. Attachment
- Adherence of bacterial cells to surface
- Reversible and dynamic process

2. Microcolony Formation
- Utilization of nutrients
- Multiplication of microbial cells
- Inhibition of locomotion

3. Matrix Formation and Biofilm maturation
- Secretion of extracellular polysaccharides
- Quorum sensing helping in maturation of biofilms
- Prevention from attack of harmful organisms

4. Dispersion
- Caused by depletion of oxygen or nutrients
- Due to presence of toxins
- Reduction in cyclic di-GMP

Fig. (1). Stepwise process of biofilm formation.

Initially, biofilm formation is instigated by a few cells attaching to a surface, and later on, it grows due to the communication of cells with other microbial species. The matrix so formed becomes extensive with the growth of biofilms (Fig. **1**). The microbial communities present in a biofilm may be of the same species or belong to different species of various genera of bacteria, fungi, and algae, depending on their ability to grow on any organic or inorganic surface serving as a nutrient source.

Biofilms provide many advantages to their microbial communities, including nutrients' availability, protection from harsh environmental conditions, easy exchange of genetic materials and communication compared to their free-living counterparts [33, 34]. Biofilm formation becomes a survival strategy during adverse environmental conditions such as starvation, desiccation, *etc.*, for bacteria and protects them from the antimicrobial mechanism exhibited by the host [35].

Studies on the structure of biofilms have shown that the extra polymeric substances comprise about 50-90% of total organic carbon. In gram-negative bacteria, polyanionic or neutral polysaccharides have been found in the biofilm [36]. Divalent cations such as calcium and magnesium provide structural integrity and binding strength to biofilm structure. The bacterial biofilms also comprise biomolecules such as polysaccharides, protein, lipids, and organic substances depending upon several factors like environmental stress, growth conditions, and microbial species. Water channels present within a biofilm separate the microbial communities and help transport nutrients, oxygen, and microorganisms from one place to another [37].

Biofilms find wide applications in the treatment of industrial wastewater, and in the disinfection of polluted sites. They are used as biofuels, biopolymers, biocontrol agents, organic manure, and in the production of nanomaterials, cosmetics, food, pharmaceuticals, *etc.* (Fig. **2**). Several studies have revealed that bacterial biofilms possess the ability to degrade pollutants generated by industries by using them as a source of carbon [38]. They play an essential role in the aquatic food chain, thus, positively contributing to maintaining ecological balance [39]. Inedible biofilm formation by bacteria takes place to get protection from predatory protozoa. The composition of biofilms is equally significant towards the particular functions performed by the bacterial communities present within them [40]. In aquatic systems, microbial aggregates are attached to sediment surfaces from mats comprising algae and phototrophic bacteria to combat environmental stress and predation. Gram-negative bacteria present in biofilms can overcome stress due to the release of membrane vesicles [41]. It has been found that under starvation stress, *Pseudomonas aeruginosa* produces predatory membrane vesicles that can kill other bacteria, thereby making nutrients available to biofilm microorganisms [42]. It has also been reported that *Escherichia coli* formed membrane vesicles able to neutralize the effect of polymyxin B [41]. In recent years, biofilms have been advocated as indicators of pollution, mainly heavy metal pollution in aquatic habitats due to the changes in the structure and physiology of biofilms in response to these pollutants. Biofilms as pollution indicators are highly valuable because of the biofilms' ability to adsorb environmental pollutants and develop very quickly in response to pollutants. The properties of biofilms that may be utilized for monitoring environmental contamination are biomass change, composition and variation in species, photosynthesis, pigments, and enzymatic activity. Additionally, molecular methods are also employed to study the variation and diversity of species present in biofilms.

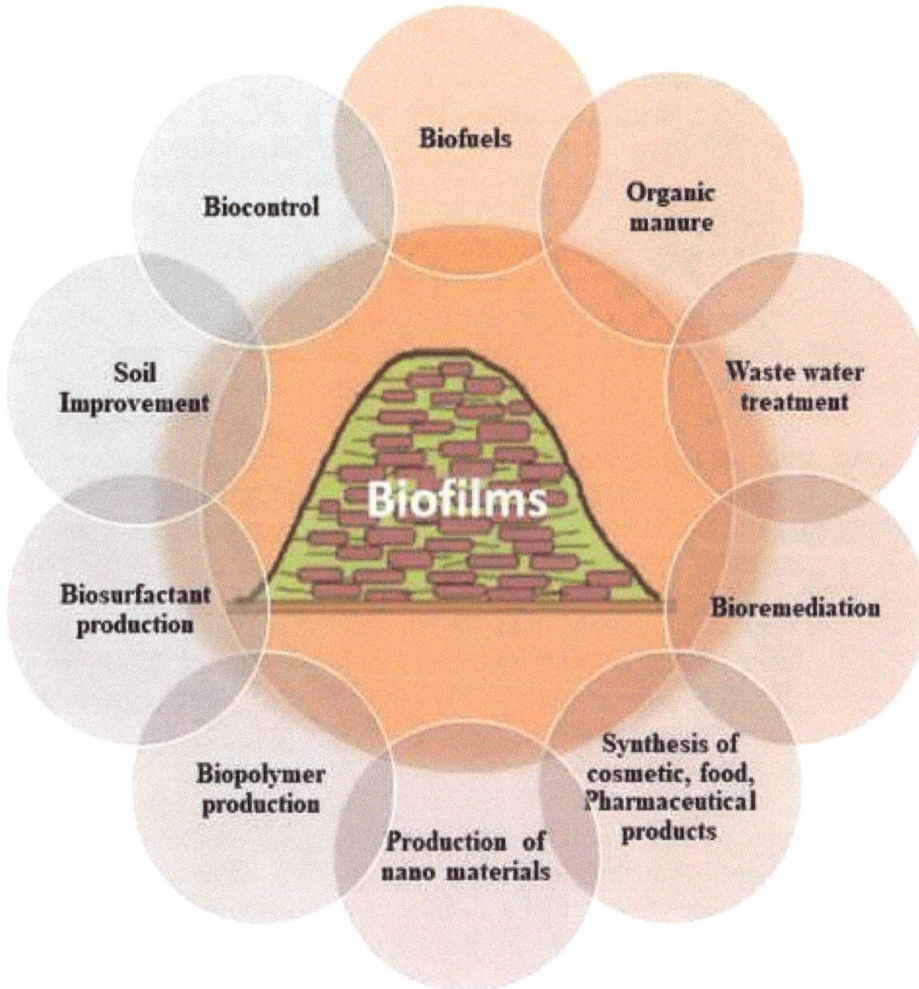

Fig. (2). Applications of biofilms.

Role of Biofilms in Bioremediation

Bioremediation using biofilms has emerged as a promising and eco-friendly approach because biofilms' microbes degrade, absorb and immobilize a large number of pollutants from various sources. Various microbes comprising biofilms can degrade a wide range of pollutants due to their different metabolic pathways.

Different microbial species in biofilms lead to differential gene expression because of variable nutrient supply, ultimately leading to a division of labor among the microorganisms of biofilms. Biofilms are known to develop in the vicinity of highly polluted sites to adapt themselves to the harsh environmental conditions of acidity, temperature, radiation, drought, predation, pollutants, *etc.*, thereby making them best suited to carry out the process of bioremediation. Various microbial responses, including chemotaxis, motility, and quorum sensing in response to various pollutants, help microbes develop biofilms [43]. Unlike free-living microbes, biofilm microorganisms are entrapped in extra polymeric substances that can immobilize various pollutants, thus facilitating bioremediation.

Moreover, free-living microorganisms are not stationary and not well adapted to environmental stress. Due to the structure of EPM, there is differential oxygen concentration in biofilms so that aerobes can grow along with anaerobes, sulfate reducers with sulfate oxidizers, and heterotrophs with nitrifiers, thereby leading to faster pollutant degradation. Due to several negatively charged groups on EPM, biofims can attract and adsorb metals like manganese, copper, lead, mercury, zinc, magnesium, iron, cadmium, nickel, *etc.*, removing them from the environment. The synthesis of EPM is known to be enhanced under nutrient-limiting conditions. Nutrient exchange and by-product elimination also become easy due to EPM.

Biofilms have been utilized in the degradation and removal of various organic and inorganic pollutants from the environment. In particular, biofilms are highly effective in removing heavy metals like cadmium, copper, chromium, and uranium from the soil, groundwater, and wastewater [44, 45]. It has been found that phosphatase activity of biofilms helps in metal precipitation. Biofilms able to degrade persistent organic compounds have been isolated and characterized. Major factors involved in the degradation of polycyclic aromatic hydrocarbons are the ability to form biofilms and cometabolism by microbial communities of biofilms. It has also been found that a surfactant-producing microbe can enhance biodegradation, as reported for the bacterial consortium of *Acinetobacter radioresistens* and *Bacillus subtilis* along with a surfactant-producing strain [46].

Organic Pollutants

Various organic pollutants from various sources have been reported to be treated using microbial biofilms Table **1**.

Table 1. Bioremediation of organic pollutants by biofilm-forming microorganisms.

Microorganisms in Biofilms	Source/Conditions	Pollutants Degraded	References
Mixed culture of herbicide-degrading bacteria	Granular activated-carbon biofilm reactor	Mecoprop (MCPP), 2,4-D	[88]
Methylosinus trichosporium	Laboratory-scale rotating drum biofilm reactor	Acid Orange 10, 14	[89]
Pseudomonas sp., *Rhodococcus* sp.	Fluidized bed biofilm reactor	2,3,4,6-tetrachlorophenol	[90]
Providencia stuartii, *Pseudomonas cepacia*	Continuous flow fixed biofilm reactor	Carbon tetrachloride	[91]
Prototheca zopfii	Rotating biological contactors	*n*-Alkanes	[92]
Alcaligenes eutrophus	Membrane feeding substrate bioreactor	Nitrate	[5]
Polaromonas sp., *Sphingomonas* sp., *Alcaligenes* sp., *Caulobacter* and *Variovorax* sp.	Biofilm grown directly in liquid medium	Pyrene, phenanthrene	[93]
Coriolus versicolor	Laboratory-scale activated sludge unit	Everzol Turquoise Blue G	[94]
Pseudomonas fluorescens	Biofilm grown in NAPLs	o-Cresol, naphthalene, phenol, 1,2,3-trimethylbenzene	[95]
Desulfitobacterium frappieri PCP-1	Anaerobic upflow sludge bed	Pentachlorophenol (PCP)	[96]
Anaerobic sludge from a swine wastewater treatment plant	Silicone tube membrane bioreactor	2-Chlorophenol	[97]
Pseudomonas putida	Rotating perforated tube biofilm reactor	2,4-Dichlorophenol	[98]
Pseudomonas putida strain Tol 1A	Hollow-fiber membrane biofilm reactor	Toluene; Para-chloronitrobenzene (p-CNB)	[6]
Dehalobium chlorocoercia DF1 and *Burkholderia xenovorans* strain LB400	Bioactive granular activated carbon	Polychlorinated biphenyls (PCBs) mixed PCB contamination	[57]
Pseudomonas otitidis	River, estuary and sea-water samples	Crude oil degradation	[99]
Candida viswanathii TH1, *Candida tropicalis* TH4, and *Trichosporon asahii* B1	Hydrocarbon-contaminated water and sediment samples in Vietnam	Phenol, naphthalene, anthracene, and pyrene	[54]

(Table 1) cont.....

Microorganisms in Biofilms	Source/Conditions	Pollutants Degraded	References
P. monteilii P26 and *Gordonia* sp. H19	Polyurethane foam	Crude oil removal	[100]
Cupriavidus necator JMP134		2,4-Dichlorophenoxyacetic acid	[101]
B. atrophaeus CN4	Oil-contaminated soil	Naphthalene	[102]
Burkholderiales, *Xanthomonadales*, *Flavobacteriales*, and *Sphingobacteriales*	Hospital wastewater	Pharmaceuticals	[7]
Rhodococcus rhodochrous BX2 1, *B. mojavensis* M1	Groundwater	Organic cyanide	[103]

Polycyclic Aromatic Hydrocarbons (PAHs)

More than hundreds of polycyclic aromatic hydrocarbons (PAH) have been known to exist in the environment, and their mixtures are the most common organic pollutants. Due to their low solubility and availability, it is challenging to remove or degrade these PAHs. Consumption of PAHs *via* the food chain leads to their bioaccumulation leading to mutagenic and carcinogenic effects. Biofilms are beneficial in bioremediation as they can increase the solubility of these PAHs (Table **1**). Degradation of PAHs by bacteria is carried out by producing monooxygenase or dioxygenase under aerobic conditions. The aromatic ring of PAHs is hydroxylated by the dioxygenase enzyme, thereby forming cis-dihydrodiolre which is further aromatized by dehydrogenase into a diol intermediate. Subsequently, the diol intermediates are broken down by dioxygenases and converted into TCA cycle intermediates [47, 48]. Additionally, bacteria also degrade PAHs with the production of trans-dihydrodiols by utilizing a cytochrome P450-mediated pathway or anaerobically under nitrate-reducing conditions [49 - 51].

It has been reported that the bacterial plasmids carry the catabolic genes along with the genes for chemotaxis for PAHs, thereby helping the bacteria to adapt to PAH pollutants [52]. Using the cometabolism phenomenon, multiple PAHs can be degraded simultaneously [53]. Biofilm-forming yeast belonging to *Candida viswanathii, Candida tropicalis,* and *Trichosporon asahii* isolated from oil-contaminated sediment and water were tested for the degradation of PAHs, including phenol, naphthalene, anthracene, and pyrene [54]. It was found that the consortium of the strains after seven days of incubation was able to degrade 90, 85, 82, and 67% of phenol, naphthalene, anthracene, and pyrene, respectively. Gas chromatography-mass spectrometry (GCMS) analysis showed that the biofilm could degrade these compounds completely after 14 days of incubation.

Polychlorinated Biphenyls (PCBs) and Dioxins

In addition to PAHs, polychlorinated biphenyls (PCBs) and dioxins are also highly toxic organic pollutants. Though these compounds are banned, they can still be found in various environments. Biofilm-forming microorganisms can dechlorinate PCBs, thus remediating various environments (Table **1**). *Chloroflexi* spp. has been shown to dechlorinate PCBs reductively under anaerobic conditions, thereby leading to less chlorinated compounds that may be degraded under aerobic conditions [55]. Biofilm community developed on Permanox as substrate showed the degradation of Aroclor, a PCB oil in contaminated soil [56]. Methods including confocal laser scanning microscopy, 16S rRNA gene amplicons, single-strand conformational polymorphism, and analysis of PCB congeners were used to monitor the biofilm for 31 days. *Bradyrhizobium* and *Herbaspirillum* were found to be the dominant members of the young biofilm, as revealed by single-strand conformational polymorphism [56].

Aroclor, a commercially used PCB, is converted into less chlorinated congeners by reductive dehalogenation anaerobically by microorganisms. These less chlorinating congeners are then dechlorinated and degraded by the aerobic process. Dehalorespiring bacterium *Dehalobium chlorocoerciais* reported being stimulated to dechlorinate PCBs by anaerobic bioaugmentation [57]. The concentration of pentachlorinated PCBs decreased to about 56% in bio-augmented mesocosms by *D. chlorocoercia* after 120 days compared to uninoculated controls where no degradation activity was observed. Biofilms containing dehalogenation bacteria are also reported to grow and develop on granular activated carbon due to bioaugmentation and activated charcoal sediment amendment [58].

Apart from PCBs, dioxins produced as by-products during various processes, including herbicide production, and incineration of various wastes, are also the most toxic organic contaminants [59]. In heavily polluted soil microcosms, about 22% degradation of dioxins in three months is reported with the use of microorganisms [60]. The microbial communities of heavily polluted sites were found to be different from the less polluted sites, indicating that the pollution caused by dioxin significantly affected the soil microbial population in terms of quantity, quality, and activity. Complete degradation of chlorinated benzoates has been reported using biofilm containing a consortium of *Burkholderia* sp. and *Pseudomonas aeruginosa* [61].

Chlorinated Ethenes

Chlorinated ethenes are also the organic pollutants causing pollution of water sources. Bacteria belonging to the genus Dehalococcoides have been found to carry out reductive dechlorination of chlorinated ethenes [62]. In an experiment to study the effect of hydrogen on *ex-situ* reductive dechlorination of chlorinated ethenes, it was found that at the end of the reactor operation, the dominant bacteria in the biofilm were *Geobacter, Dehalococcoides,* and *Homoacetogens* [63]. It was observed that trichloroethene was reductively dechlorinated to ethene with 95% efficiency when pH-increase effects of methanogenic and homoacetogenic bacteria were managed. In order to improve reductive dechlorination in trichloroethene polluted aquifer using *Dehalococcoides ethenogenes* with the replacement of electron donor in an experiment, it was observed that the electron donor led to the stimulation of cometabolism by methane-oxidizing *Methylosinus trichosporium* [64]. Reductive and oxidative co-metabolic degradation of polychlorinated ethenes has also been stimulated by electrochemical processes that originate from an anode-cathode bio-electro setup [65].

Pharmaceutical and Personal Care Products (PPCPs)

Many pharmaceutical and personal care products falling under xenobiotics are not regulated and are discharged into the water as industrial effluent. An industrially used antimicrobial compound Triclosan is converted into water dioxin congeners and negatively affects biofilm-forming microorganisms able to degrade dioxins [66, 67]. In sewage treatment plants, the biodegradation of triclosan is still inadequate, leading to its entry into agricultural land through sludge used as a fertilizer [68]. The pharmaceutical ibuprofen and its two metabolites, hydroxy–and carboxy–ibuprofen, are reported to be degraded by a river biofilm bioreactor [69]. Membrane biofilm reactors have been proven to be more beneficial in removing PPCPs than traditional activated sludge [70]. Biofilm-forming bacterial communities able to degrade these compounds are found to be less active in winters, as indicated by increased concentrations of these pharmaceuticals during winters in wastewater effluents. In a study, degradation of ibuprofen, clofibric acid, and diclofenac in pilot sewage plants and biofilm reactors were tested [71]. The results indicated that the concentrations of all three pharmaceuticals were decreased to about 60 and 80% of their initial concentration in the anoxic biofilm reactor. The efficiency of *in situ* bioremediation of PPCPs was tested using wastewater sludge as inoculum for biofilm formation on sand filters [72]. Out of 14 PPCPs, ten were found to be thoroughly degraded, indicating the significant effect of biofilms in reducing the effect of these pharmaceuticals. Biofilm-forming Burkholderiales, Xanthomonadales, Flavobacteriales, and Sphingobacteriales have recently been reported to degrade the pharmaceuticals in hospital wastewater [7].

Inorganic Pollutants

Heavy Metals

The presence of heavy metals in terrestrial and aquatic habitats derived from various industries is one of the main environmental issues leading to the prioritization of their removal from the environment. Many biofilm-forming microorganisms are reported to detoxify the effect of several heavy metals like cadmium, mercury, cobalt, nickel, copper, arsenic, lead, chromium, zinc, *etc.* (Table **2**). Consortium cultures in biofilms have been shown to absorb more mercury than single culture biofilms during treating mercury-contaminated wastewater in packed-bed bioreactors [73]. Diverse microorganisms in biofilms help in a better establishment in different habitats of varying stress conditions leading to their better ability in degradation. During the treatment of industrial wastewater with a high heavy metal concentration in a moving bed sand filter, bacterial biofilms served as biosorptive material and lowered the pH leading to the precipitation of heavy metals [74]. Mixed cultures in biofilms comprising sulfate-reducing bacteria reported removal of about 98% copper, zinc and nickel, 82% iron, and 78% arsenic [75].

Table 2. Biodegradation of heavy metals by biofilm-forming bacteria.

Biofilm Forming Microorganisms	Source	Pollutants Degraded	References
Cycloclasticus sp.	Soil	PAH	[104]
Bacillus circulans, Bacillus megaterium	common effluent treatment plant (CETP) at Unnao (26.48N, 80.43E), Uttar Pradesh	Cr	[105]
Staphylococcus aureus	Marine and freshwater environments	Chromate	[106]
Achromobacter sp.	Lead-contaminated industrial site in Australia	Hg	[107]
Gloeophyllum sepiarium	Cr-contaminates soil in Bokaro, Jharkhand	Cr (VI)	[108]
Aspergillus fumigatus	Soil	Pb	[109]
Sporosarcinagin sengisoli	Arsenic-contaminated soil of Urumqi	As (III)	[110]
Geobacillus thermolevorans subsp. *stromboliensis, Geobacillustoebii* subsp. *decanicus*	Soil	Cd, Cu, Co, Mn	[111]

(Table 2) cont.....

Biofilm Forming Microorganisms	Source	Pollutants Degraded	References
Pseudomonas aeruginosa	Soil	Inorganic mercury	[112]
Pseudomonas putida, Bacillus subtilis	Tannery effluent contaminated sites of Palar river basin in Vellore District, Tamil Nadu	Cr (VI)	[113]
Bacillus subtilis	ash dyke sample of four thermal power plants of Chhattisgarh	Cd, Cr, Hg, Pb, Ni, Co, Hg, Cu, Mn	[114]
Pseudomonas aeruginosa	Marine bacterium	Pb	[8]

Biofilm-forming cadmium-resistant *Aeromonas* sp. has been reported to bio-augment cadmium in the rhizosphere of Vetiveriazizanioides [76]. Recently, acid phosphatase produced by *Staphylococcus aureus* biofilm has been reported to remediate uranium pollution [77]. It was observed that the biofilm could remediate about 47% of the 10 ppm uranium, and both enzymatic activity and biosorption by *S. aureus* biofilms carried out the process of uranium remediation. The addition of phosphate substrate leads to the enhancement in remediation [77].

Characterization and Identification of Microorganisms in Biofilms

The efficiency of biofilms in degrading the pollutants depends upon the type of microorganisms, so it becomes necessary to characterize and identify the organisms in a biofilm. Various traditional, molecular, spectroscopic, and microsensors approaches have been used to identify the microorganisms in a biofilm. The traditional methods, including light microscopy or culture-based techniques, are unreliable as only 1-5% of the microbial communities are cultivable. Recently, various molecular techniques have been successfully employed to identify biofilm microbes. The techniques used include real-time polymerase chain reaction, denaturing gradient gel electrophoresis (DGGE), fluorescent *in situ* hybridization (FISH), temperature gradient gel electrophoresis (TGGE), denaturing high-performance liquid chromatography (DHPLC), immunoassays, metagenomic sequencing. The bacteria that purified wastewater were identified as *Enterobacter agglomerans*, *Pantoea agglomerans,* and *Cronobater sakazakii* using DGGE in many biofilms samples [78]. In the same study, the bacterial community showed variable UV resistance while studying the inactivation of bacteria with (UV)-C254. Four phenol degrading and biofilm-forming bacteria were identified as *Rhodococcus equi, Enterobacter mori, Micrococcus* sp., and *Bacillus* sp. based on phenotypic and genotypic characters [79].

In addition to molecular methods, spectroscopic methods including fluorescence spectrometry, atomic absorption spectrometry, and spectrophotometry; surface and interface-characterizing techniques like atomic adsorption microscopy, X-ray microscopy, scanning electron microscopy, confocal laser scanning microscopy, nuclear magnetic resonance; and electrochemical and fiber-optic microsensors are also used for characterization [80 - 82]. Knowledge regarding the type of microbes present in a biofilm helps evaluate the degradation process.

Limitations

Through bioremediation using biofilms is a practical and cost-effective technology to remove toxic pollutants from the environment, but it suffers from some limitations and disadvantages, including the limited number of microorganisms with the potential to degrade pollutants effectively; slow rate of degradation compared to chemical methods; adsorption of pollutants not accessible to microorganisms; non-availability of nutrients, electron donors or acceptors, water activity, *etc.* leading to poor microbial growth. Additionally, microorganisms are not degradable by many pollutants, including some halogenated compounds and plastics. Bioremediation technology is applicable at the low level of contaminants/pollutants. Many biodegradation processes also sometimes produce toxic products, further aggravating the problem of pollution generation. The efficacy of remediation by microorganisms is also determined by the microbial accessibility of pollutants not enclosed by other materials. For biofilm-mediated bioremediation's successful outcome, an interdisciplinary approach and technical expertise are highly required.

The lack of native biofilm-forming microbial populations able to degrade recalcitrant and persistent pollutants like PCBs also limits the process. Additionally, bioremediation becomes slow as the dechlorination process is less energy-yielding, leading to a microbial population sensitive to various environmental stresses [83]. Moreover, only very few bacteria are reported to carry out the reductive dechlorination of congeners with 4-5 chlorines so that further process of complete degradation by aerobic microorganisms can be carried out [62, 84, 85]. Likewise, many bioremediation processes are deficient in electron acceptors. During the remediation of metals by *Geobacter* spp., the limiting factor is Fe (III) as an electron acceptor [86]. Similarly, microarray analysis has shown that phosphate is a limiting factor for *Geobacter sulfureducens* [87].

CONCLUSION

Biofilm mediation bioremediation is a promising and eco-friendly approach because the microbes of biofilms degrade, absorb and immobilize a large number of pollutants from various sources. Biofilms containing a different group of microorganisms can degrade a wide range of pollutants due to their different metabolic pathways. Compared to free-living microorganisms, biofilm-forming microbes are entrapped in extra polymeric substances, thus able to immobilize various pollutants and facilitate the bioremediation process. Considering a few limitations of the technology, improved microbial strains with the ability to prefer pollutants as nutrient substrates are being tested to make the bioremediation more effective. Genetically modified microorganisms able to degrade a large number of pollutants are also being designed. Enhancing the chemotaxis of microbes towards pollutants may further improve the bioremediation by biofilms; however, the use of genetically engineered microorganisms in natural environments is controversial and debatable. The proteins secreted by microbes in the biofilm matrix may also be engineered to further improve the process.

CONSENT FOR PUBLICATION

Not applicable.

CONFLICT OF INTEREST

The authors declare that there is no conflict of interest.

ACKNOWLEDGEMENTS

The author acknowledges Punjab Agricultural University, Punjab, for providing the necessary facilities and support in formulating the manuscript.

REFERENCES

[1] Edwards SJ, Kjellerup BV. Applications of biofilms in bioremediation and biotransformation of persistent organic pollutants, pharmaceuticals/personal care products, and heavy metals. Appl Microbiol Biotechnol 2013; 97(23): 9909-21.
[http://dx.doi.org/10.1007/s00253-013-5216-z] [PMID: 24150788]

[2] Vu B, Chen M, Crawford R, Ivanova E. Bacterial extracellular polysaccharides involved in biofilm formation. Molecules 2009; 14(7): 2535-54.
[http://dx.doi.org/10.3390/molecules14072535] [PMID: 19633622]

[3] Banerjee A. Toxic effect and bioremediation of oil contamination in algal perspective. In. Kumar V, Saxena G, Shah MP, Eds. Bioremediation for Environmental Sustainability. Elsevier, 2020; 283-98.

[4] Boles BR, Thoendel M, Singh PK. Self-generated diversity produces "insurance effects" in biofilm communities. Proc Natl Acad Sci USA 2004; 101(47): 16630-5.
[http://dx.doi.org/10.1073/pnas.0407460101] [PMID: 15546998]

[5] Ong SA, Ho LN, Wong YS, Raman K. Performance and kinetic study on bioremediation of diazo dye (Reactive Black 5) in wastewater using spent GAC–biofilm sequencing batch reactor. Water Air Soil Pollut 2012; 223(4): 1615-23.
[http://dx.doi.org/10.1007/s11270-011-0969-4]

[6] Tribelli PM, Di Martino C, López NI, Raiger Iustman LJ. Biofilm lifestyle enhances diesel

bioremediation and biosurfactant production in the Antarctic polyhydroxyalkanoate producer *Pseudomonas extremaustralis.* Biodegradation 2012; 23(5): 645-51.
[http://dx.doi.org/10.1007/s10532-012-9540-2] [PMID: 22302594]

[7]   Torresi E, Gülay A, Polesel F, *et al.* Reactor staging influences microbial community composition and diversity of denitrifying MBBRs- Implications on pharmaceutical removal. Water Res 2018; 138: 333-45.
[http://dx.doi.org/10.1016/j.watres.2018.03.014] [PMID: 29635164]

[8]   Kumari S, Das S. Expression of metallothionein encoding gene bmtA in biofilm-forming marine bacterium *Pseudomonas aeruginosa* N6P6 and understanding its involvement in Pb(II) resistance and bioremediation. Environ Sci Pollut Res Int 2019; 26(28): 28763-74.
[http://dx.doi.org/10.1007/s11356-019-05916-2] [PMID: 31376126]

[9]   Accinelli C, Saccà ML, Mencarelli M, Vicari A. Application of bioplastic moving bed biofilm carriers for the removal of synthetic pollutants from wastewater. Bioresour Technol 2012; 120: 180-6.
[http://dx.doi.org/10.1016/j.biortech.2012.06.056] [PMID: 22797083]

[10]   Stokes JD, Paton GI, Semple KT. Behaviour and assessment of bioavailability of organic contaminants in soil: relevance for risk assessment and remediation. Soil Use Manage 2005; 21(1): 475-86.
[http://dx.doi.org/10.1079/SUM2005347]

[11]   Valentin L, Nousiainen A, Mikkonen A. Introduction to organic contaminants in soil. concepts and risks. In. Vicent T, Caminal G, Eljarrat E, Barceló D, Eds. Emerging Organic Contaminants in Sludges.Springer, Berlin, Heidelberg, 2013; 1-29.
[http://dx.doi.org/10.1007/698_2012_208]

[12]   Andreu V, Picó Y. Determination of pesticides and their degradation products in soil: critical review and comparison of methods. Trends Analyt Chem 2004; 23(10-11): 772-89.
[http://dx.doi.org/10.1016/j.trac.2004.07.008]

[13]   Bermúdez-Couso A, Arias-Estévez M, Nóvoa-Muñoz JC, López-Periago E, Soto-González B, Simal-Gándara J. Seasonal distributions of fungicides in soils and sediments of a small river basin partially devoted to vineyards. Water Res 2007; 41(19): 4515-25.
[http://dx.doi.org/10.1016/j.watres.2007.06.029] [PMID: 17624393]

[14]   Gennadiev AN, Tsibart AS. Pyrogenic polycyclic aromatic hydrocarbons in soils of reserved and anthropogenically modified areas: Factors and features of accumulation. Eurasian Soil Sci 2013; 46(1): 28-36.
[http://dx.doi.org/10.1134/S106422931301002X]

[15]   Wang HJ, Chen HP. Understanding the recent trend of haze pollution in eastern China: roles of climate change. Atmos Chem Phys 2016; 16(6): 4205-11.
[http://dx.doi.org/10.5194/acp-16-4205-2016]

[16]   Lièvremont D, Bertin PN, Lett MC. Arsenic in contaminated waters: Biogeochemical cycle, microbial metabolism and biotreatment processes. Biochimie 2009; 91(10): 1229-37.
[http://dx.doi.org/10.1016/j.biochi.2009.06.016] [PMID: 19567262]

[17]   Kumar A, Bisht BS, Joshi VD, Dhewa T. Review on bioremediation of polluted environment. a management tool. Int J Environ Sci 2011; 1(6): 1079-93.

[18]   Mueller JG, Cerniglia CE, Pritchard PH. Bioremediation of environments contaminated by polycyclic aromatic hydrocarbons. Biotechnol Res Series 1996; 6: 125-94.
[http://dx.doi.org/10.1017/CBO9780511608414.007]

[19]   Hou D, O'Connor D, Igalavithana AD, *et al.* Metal contamination and bioremediation of agricultural soils for food safety and sustainability. Nat Rev Earth Environ 2020; 1(7): 366-81.
[http://dx.doi.org/10.1038/s43017-020-0061-y]

[20]   Hou J, Lin D, White JC, Gardea-Torresdey JL, Xing B. Joint nanotoxicology assessment provides a new strategy for developing nanoenabled bioremediation technologies. Environ Sci Technol 2019;

53(14): 7927-9.
[http://dx.doi.org/10.1021/acs.est.9b03593] [PMID: 31269395]

[21]   Kallmeyer J, Pockalny R, Adhikari RR, Smith DC, D'Hondt S. Global distribution of microbial abundance and biomass in subseafloor sediment. Proc Natl Acad Sci USA 2012; 109(40): 16213-6.
[http://dx.doi.org/10.1073/pnas.1203849109] [PMID: 22927371]

[22]   Serna-Chavez HM, Fierer N, van Bodegom PM. Global drivers and patterns of microbial abundance in soil. Glob Ecol Biogeogr 2013; 22(10): 1162-72.
[http://dx.doi.org/10.1111/geb.12070]

[23]   Anantharaman K, Brown CT, Hug LA, *et al.* Thousands of microbial genomes shed light on interconnected biogeochemical processes in an aquifer system. Nat Commun 2016; 7(1): 13219.
[http://dx.doi.org/10.1038/ncomms13219] [PMID: 27774985]

[24]   Frutos FJG, Pérez R, Escolano O, *et al.* Remediation trials for hydrocarbon-contaminated sludge from a soil washing process: Evaluation of bioremediation technologies. J Hazard Mater 2012; 199-200: 262-71.
[http://dx.doi.org/10.1016/j.jhazmat.2011.11.017] [PMID: 22118850]

[25]   Smith E, Thavamani P, Ramadass K, Naidu R, Srivastava P, Megharaj M. Remediation trials for hydrocarbon-contaminated soils in arid environments: Evaluation of bioslurry and biopiling techniques. Int Biodeterior Biodegradation 2015; 101: 56-65.
[http://dx.doi.org/10.1016/j.ibiod.2015.03.029]

[26]   Azubuike CC, Chikere CB, Okpokwasili GC. Bioremediation techniques–classification based on site of application: principles, advantages, limitations and prospects. World J Microbiol Biotechnol 2016; 32(11): 180.
[http://dx.doi.org/10.1007/s11274-016-2137-x] [PMID: 27638318]

[27]   Scow KM, Hicks KA. Natural attenuation and enhanced bioremediation of organic contaminants in groundwater. Curr Opin Biotechnol 2005; 16(3): 246-53.
[http://dx.doi.org/10.1016/j.copbio.2005.03.009] [PMID: 15961025]

[28]   García-Delgado C, Alfaro-Barta I, Eymar E. Combination of biochar amendment and mycoremediation for polycyclic aromatic hydrocarbons immobilization and biodegradation in creosote-contaminated soil. J Hazard Mater 2015; 285: 259-66.
[http://dx.doi.org/10.1016/j.jhazmat.2014.12.002] [PMID: 25506817]

[29]   Silva-Castro GA, Uad I, Gónzalez-López J, Fandiño CG, Toledo FL, Calvo C. Application of selected microbial consortia combined with inorganic and oleophilic fertilizers to recuperate oil-polluted soil using land farming technology. Clean Technol Environ Policy 2012; 14(4): 719-26.
[http://dx.doi.org/10.1007/s10098-011-0439-0]

[30]   Bhattacharya M, Guchhait S, Biswas D, Datta S. Waste lubricating oil removal in a batch reactor by mixed bacterial consortium: a kinetic study. Bioprocess Biosyst Eng 2015; 38(11): 2095-106.
[http://dx.doi.org/10.1007/s00449-015-1449-9] [PMID: 26271337]

[31]   O'Connor D, Hou D, Ok YS, Lanphear BP. The effects of iniquitous lead exposure on health. Nat Sustain 2020; 3(2): 77-9.
[http://dx.doi.org/10.1038/s41893-020-0475-z]

[32]   Kumar M, Jaiswal S, Sodhi KK, *et al.* Antibiotics bioremediation: Perspectives on its ecotoxicity and resistance. Environ Int 2019; 124: 448-61.
[http://dx.doi.org/10.1016/j.envint.2018.12.065] [PMID: 30684803]

[33]   Regenberg B, Hanghøj KE, Andersen KS, Boomsma JJ. Clonal yeast biofilms can reap competitive advantages through cell differentiation without being obligatorily multicellular. Proc Royal Soc B: BiolSci 2016; 16; 283(1842): 20161303.

[34]   Lohse MB, Gulati M, Johnson AD, Nobile CJ. Development and regulation of single- and multi-species *Candida albicans* biofilms. Nat Rev Microbiol 2018; 16(1): 19-31.

[http://dx.doi.org/10.1038/nrmicro.2017.107] [PMID: 29062072]

[35]   Banerjee D, Shivapriya PM, Gautam PK, Misra K, Sahoo AK, Samanta SK. A review on basic biology of bacterial biofilm infections and their treatments by nanotechnology-based approaches. Proc Natl Acad Sci, India, Sect B Biol Sci 2020; 90(2): 243-59.
[http://dx.doi.org/10.1007/s40011-018-01065-7]

[36]   Sharma G, Karnwal A. [36] Sharma G, Karnwal A. Biological Strategies Against Biofilms. In Microbial Biotechnology. Basic Research and Applications. Springer, Singapore, 2020; 205-32.
[http://dx.doi.org/10.1007/978-981-15-2817-0_9]

[37]   Vasudevan R. Biofilms. microbial cities of scientific significance. J Microbiol Exp 2014; 1(3): 00014.
[http://dx.doi.org/10.15406/jmen.2014.01.00014]

[38]   Sgountzos IN, Pavlou S, Paraskeva CA, Payatakes AC. Growth kinetics of *Pseudomonas fluorescens* in sand beds during biodegradation of phenol. Biochem Eng J 2006; 30(2): 164-73.
[http://dx.doi.org/10.1016/j.bej.2006.03.005]

[39]   Jung JH, Lee SS, Shinkai S, Iwaura R, Shimizu T. Novel silica nanotubes using a library of carbohydrate gel assemblies as templates for sol-gel transcription in binary systems. Bull Korean Chem Soc 2004; 25(1): 63-8.
[http://dx.doi.org/10.5012/bkcs.2004.25.1.063]

[40]   Costerton JW, Lewandowski Z, Caldwell DE, Korber DR, Lappin-Scott HM. Microbial biofilms. Annu Rev Microbiol 1995; 49(1): 711-45.
[http://dx.doi.org/10.1146/annurev.mi.49.100195.003431] [PMID: 8561477]

[41]   Manning AJ, Kuehn MJ. Functional advantages conferred by extracellular prokaryotic membrane vesicles. J Mol Microbiol Biotechnol 2013; 23(1-2): 131-41.
[PMID: 23615201]

[42]   Beveridge TJ, Makin SA, Kadurugamuwa JL, Li Z. Interactions between biofilms and the environment. FEMS Microbiol Rev 1997; 20(3-4): 291-303.
[http://dx.doi.org/10.1111/j.1574-6976.1997.tb00315.x] [PMID: 9299708]

[43]   Mangwani N, Kumari S, Das S. Bacterial biofilms and quorum sensing: fidelity in bioremediation technology. Biotechnol Genet Eng Rev 2016; 32(1-2): 43-73.
[http://dx.doi.org/10.1080/02648725.2016.1196554] [PMID: 27320224]

[44]   Owlad M, Aroua MK, Daud WAW, Baroutian S. Removal of hexavalent chromium-contaminated water and wastewater. a review. Water Air Soil Pollut 2009; 200(1-4): 59-77.
[http://dx.doi.org/10.1007/s11270-008-9893-7]

[45]   Igiri BE, Okoduwa SI, Idoko GO, Akabuogu EP, Adeyi AO, Ejiogu IK. Toxicity and bioremediation of heavy metals contaminated ecosystem from tannery wastewater: a review. J Toxicol 2018; 2568038.
[http://dx.doi.org/10.1155/2018/2568038]

[46]   Mnif I, Sahnoun R, Ellouz-Chaabouni S, Ghribi D. Application of bacterial biosurfactants for enhanced removal and biodegradation of diesel oil in soil using a newly isolated consortium. Process Saf Environ Prot 2017; 109: 72-81.
[http://dx.doi.org/10.1016/j.psep.2017.02.002]

[47]   Mallick S, Chakraborty J, Dutta TK. Role of oxygenases in guiding diverse metabolic pathways in the bacterial degradation of low-molecular-weight polycyclic aromatic hydrocarbons: A review. Crit Rev Microbiol 2011; 37(1): 64-90.
[http://dx.doi.org/10.3109/1040841X.2010.512268] [PMID: 20846026]

[48]   Cerniglia CE. Biodegradation of polycyclic aromatic hydrocarbons. Biodegradation 1992; 3(2-3): 351-68.
[http://dx.doi.org/10.1007/BF00129093]

[49]   Moody JD, Freeman JP, Fu PP, Cerniglia CE. Degradation of Benzo[*a*]pyrene by *Mycobacterium*

vanbaalenii PYR-1. Appl Environ Microbiol 2004; 70(1): 340-5.
[http://dx.doi.org/10.1128/AEM.70.1.340-345.2004] [PMID: 14711661]

[50] Foght J. Anaerobic biodegradation of aromatic hydrocarbons: pathways and prospects. Microbial Physiology 2008; 15(2-3): 93-120.
[http://dx.doi.org/10.1159/000121324] [PMID: 18685265]

[51] Carmona M, Zamarro MT, Blázquez B, *et al.* Anaerobic catabolism of aromatic compounds: a genetic and genomic view. Microbiol Mol Biol Rev 2009; 73(1): 71-133.
[http://dx.doi.org/10.1128/MMBR.00021-08] [PMID: 19258534]

[52] Grimm AC, Harwood CS. Chemotaxis of *Pseudomonas* spp. to the polyaromatic hydrocarbon naphthalene. Appl Environ Microbiol 1997; 63(10): 4111-5.
[http://dx.doi.org/10.1128/aem.63.10.4111-4115.1997] [PMID: 9327579]

[53] Plósz BG, Benedetti L, Daigger GT, *et al.* Modelling micro-pollutant fate in wastewater collection and treatment systems: status and challenges. Water Sci Technol 2013; 67(1): 1-15.
[http://dx.doi.org/10.2166/wst.2012.562] [PMID: 23128615]

[54] Nhi Cong LT, Ngoc Mai CT, Thanh VT, Nga LP, Minh NN. Application of a biofilm formed by a mixture of yeasts isolated in Vietnam to degrade aromatic hydrocarbon polluted wastewater collected from petroleum storage. Water Sci Technol 2014; 70(2): 329-36.
[http://dx.doi.org/10.2166/wst.2014.233] [PMID: 25051481]

[55] Fagervold SK, Watts JEM, May HD, Sowers KR. Sequential reductive dechlorination of meta-chlorinated polychlorinated biphenyl congeners in sediment microcosms by two different Chloroflexi phylotypes. Appl Environ Microbiol 2005; 71(12): 8085-90.
[http://dx.doi.org/10.1128/AEM.71.12.8085-8090.2005] [PMID: 16332789]

[56] Macedo AJ, Kuhlicke U, Neu TR, Timmis KN, Abraham WR. Three stages of a biofilm community developing at the liquid-liquid interface between polychlorinated biphenyls and water. Appl Environ Microbiol 2005; 71(11): 7301-9.
[http://dx.doi.org/10.1128/AEM.71.11.7301-7309.2005] [PMID: 16269772]

[57] Payne RB, May HD, Sowers KR. Enhanced reductive dechlorination of polychlorinated biphenyl impacted sediment by bioaugmentation with a dehalorespiring bacterium. Environ Sci Technol 2011; 45(20): 8772-9.
[http://dx.doi.org/10.1021/es201553c] [PMID: 21902247]

[58] Mercier A, Wille G, Michel C, *et al.* Biofilm formation *vs.* PCB adsorption on granular activated carbon in PCB-contaminated aquatic sediment. J Soils Sediments 2013; 13(4): 793-800.
[http://dx.doi.org/10.1007/s11368-012-0647-1]

[59] Lohman K, Seigneur C. Atmospheric fate and transport of dioxins: local impacts. Chemosphere 2001; 45(2): 161-71.
[http://dx.doi.org/10.1016/S0045-6535(00)00559-2] [PMID: 11572608]

[60] Hiraishi A, Miyakoda H, Lim BR, Hu HY, Fujie K, Suzuki J. Toward the bioremediation of dioxin-polluted soil: structural and functional analyses of *in situ* microbial populations by quinone profiling and culture-dependent methods. Appl Microbiol Biotechnol 2001; 57(1-2): 248-56.
[http://dx.doi.org/10.1007/s002530100751] [PMID: 11693929]

[61] Yoshida S, Ogawa N, Fujii T, Tsushima S. Enhanced biofilm formation and 3-chlorobenzoate degrading activity by the bacterial consortium of *Burkholderia* sp. NK8 and *Pseudomonas aeruginosa* PAO1. J Appl Microbiol 2009; 106(3): 790-800.
[http://dx.doi.org/10.1111/j.1365-2672.2008.04027.x] [PMID: 19191976]

[62] Löffler FE, Ritalahti KM, Zinder SH. Dehalococcoides and reductive dechlorination of chlorinated solvents. In. Stroo H, Leeson A, Ward C, Eds. Bioaugmentation for groundwater remediation, vol. 5. Springer, New York, 2013; Vol. 5: 39-88.
[http://dx.doi.org/10.1007/978-1-4614-4115-1_2]

[63] Ziv-El MC, Rittmann BE. Systematic evaluation of nitrate and perchlorate bioreduction kinetics in groundwater using a hydrogen-based membrane biofilm reactor. Water Res 2009; 43(1): 173-81.
[http://dx.doi.org/10.1016/j.watres.2008.09.035] [PMID: 18951606]

[64] Conrad ME, Brodie EL, Radtke CW, *et al.* Field evidence for co-metabolism of trichloroethene stimulated by addition of electron donor to groundwater. Environ Sci Technol 2010; 44(12): 4697-704.
[http://dx.doi.org/10.1021/es903535j] [PMID: 20476753]

[65] Lohner ST, Becker D, Mangold KM, Tiehm A. Sequential reductive and oxidative biodegradation of chloroethenes stimulated in a coupled bioelectro-process. Environ Sci Technol 2011; 45(15): 6491-7.
[http://dx.doi.org/10.1021/es200801r] [PMID: 21678913]

[66] Latch DE, Packer JL, Arnold WA, McNeill K. Photochemical conversion of triclosan to 2,8-dichlorodibenzo-p-dioxin in aqueous solution. J Photochem Photobiol Chem 2003; 158(1): 63-6.
[http://dx.doi.org/10.1016/S1010-6030(03)00103-5]

[67] Buth JM, Grandbois M, Vikesland PJ, McNeill K, Arnold WA. Aquatic photochemistry of chlorinated triclosan derivatives: potential source of polychlorodibenzo-p-dioxins. Environ Toxicol Chem 2009; 28(12): 2555-63.
[http://dx.doi.org/10.1897/08-490.1] [PMID: 19908930]

[68] Heidler J, Halden RU. Mass balance assessment of triclosan removal during conventional sewage treatment. Chemosphere 2007; 66(2): 362-9.
[http://dx.doi.org/10.1016/j.chemosphere.2006.04.066] [PMID: 16766013]

[69] Winkler M, Lawrence JR, Neu TR. Selective degradation of ibuprofen and clofibric acid in two model river biofilm systems. Water Res 2001; 35(13): 3197-205.
[http://dx.doi.org/10.1016/S0043-1354(01)00026-4] [PMID: 11487117]

[70] Sui Q, Huang J, Deng S, Chen W, Yu G. Seasonal variation in the occurrence and removal of pharmaceuticals and personal care products in different biological wastewater treatment processes. Environ Sci Technol 2011; 45(8): 3341-8.
[http://dx.doi.org/10.1021/es200248d] [PMID: 21428396]

[71] Zwiener C, Frimmel F. Short-term tests with a pilot sewage plant and biofilm reactors for the biological degradation of the pharmaceutical compounds clofibric acid, ibuprofen, and diclofenac. Sci Total Environ 2003; 309(1-3): 201-11.
[http://dx.doi.org/10.1016/S0048-9697(03)00002-0] [PMID: 12798104]

[72] Onesios KM, Bouwer EJ. Biological removal of pharmaceuticals and personal care products during laboratory soil aquifer treatment simulation with different primary substrate concentrations. Water Res 2012; 46(7): 2365-75.
[http://dx.doi.org/10.1016/j.watres.2012.02.001] [PMID: 22374299]

[73] von Canstein H, Kelly S, Li Y, Wagner-Döbler I. Species diversity improves the efficiency of mercury-reducing biofilms under changing environmental conditions. Appl Environ Microbiol 2002; 68(6): 2829-37.
[http://dx.doi.org/10.1128/AEM.68.6.2829-2837.2002] [PMID: 12039739]

[74] Diels L, Spaans PH, Van Roy S, *et al.* Heavy metals removal by sand filters inoculated with metal sorbing and precipitating bacteria. Hydrometallurgy 2003; 71(1-2): 235-41.
[http://dx.doi.org/10.1016/S0304-386X(03)00161-0]

[75] Jong T, Parry DL. Removal of sulfate and heavy metals by sulfate reducing bacteria in short-term bench scale upflow anaerobic packed bed reactor runs. Water Res 2003; 37(14): 3379-89.
[http://dx.doi.org/10.1016/S0043-1354(03)00165-9] [PMID: 12834731]

[76] Itusha A, Osborne WJ, Vaithilingam M. Enhanced uptake of Cd by biofilm forming Cd resistant plant growth promoting bacteria bioaugmented to the rhizosphere of *Vetiveria zizanioides*. Int J Phytoremediation 2019; 21(5): 487-95.
[http://dx.doi.org/10.1080/15226514.2018.1537245] [PMID: 30648408]

[77]   Shukla SK, Hariharan S, Rao TS. Uranium bioremediation by acid phosphatase activity of *Staphylococcus aureus* biofilms: Can a foe turn a friend? J Hazard Mater 2020; 384: 121316.
[http://dx.doi.org/10.1016/j.jhazmat.2019.121316] [PMID: 31607578]

[78]   Yousra Turki , Mehri I, Lajnef R, *et al.* Biofilms in bioremediation and wastewater treatment: characterization of bacterial community structure and diversity during seasons in municipal wastewater treatment process. Environ Sci Pollut Res Int 2017; 24(4): 3519-30.
[http://dx.doi.org/10.1007/s11356-016-8090-2] [PMID: 27878485]

[79]   Khusnuryani A, Martani E, Wibawa T, Widada J. Molecular identification of phenol-degrading and biofilm-forming bacteria from wastewater and peat soil. Indones J Biotechnol 2016; 19(2): 99-110.
[http://dx.doi.org/10.22146/ijbiotech.9299]

[80]   Chakraborty J, Das S. Application of spectroscopic techniques for monitoring microbial diversity and bioremediation. Appl Spectrosc Rev 2017; 52(1): 1-38.
[http://dx.doi.org/10.1080/05704928.2016.1199028]

[81]   Wolf G, Crespo JG, Reis MAM. Optical and spectroscopic methods for biofilm examination and monitoring. Rev Environ Sci Biotechnol 2002; 1(3): 227-51.
[http://dx.doi.org/10.1023/A:1021238630092]

[82]   Wilson C, Lukowicz R, Merchant S, *et al.* Quantitative and qualitative assessment methods for biofilm growth. A mini-review. Res Rev J Eng Technol 2017; 6(4): 1-42.
[PMID: 30214915]

[83]   May HD, Cutter LA, Miller GS, Milliken CE, Watts JEM, Sowers KR. Stimulatory and inhibitory effects of organohalides on the dehalogenating activities of PCB-dechlorinating bacterium o-17. Environ Sci Technol 2006; 40(18): 5704-9.
[http://dx.doi.org/10.1021/es052521y] [PMID: 17007129]

[84]   Fagervold SK, May HD, Sowers KR. Microbial reductive dechlorination of aroclor 1260 in Baltimore harbor sediment microcosms is catalyzed by three phylotypes within the phylum Chloroflexi. Appl Environ Microbiol 2007; 73(9): 3009-18.
[http://dx.doi.org/10.1128/AEM.02958-06] [PMID: 17351091]

[85]   Kjellerup BV, Sun X, Ghosh U, May HD, Sowers KR. Site-specific microbial communities in three PCB-impacted sediments are associated with different *in situ* dechlorinating activities. Environ Microbiol 2008; 10(5): 1296-309.
[http://dx.doi.org/10.1111/j.1462-2920.2007.01543.x] [PMID: 18312399]

[86]   O'Neil RA, Holmes DE, Coppi MV, *et al.* Gene transcript analysis of assimilatory iron limitation in Geobacteraceae during groundwater bioremediation. Environ Microbiol 2008; 10(5): 1218-30.
[http://dx.doi.org/10.1111/j.1462-2920.2007.01537.x] [PMID: 18279349]

[87]   N'Guessan AL, Elifantz H, Nevin KP, *et al.* Molecular analysis of phosphate limitation in Geobacteraceae during the bioremediation of a uranium-contaminated aquifer. ISME J 2010; 4(2): 253-66.
[http://dx.doi.org/10.1038/ismej.2009.115] [PMID: 20010635]

[88]   Kye-Heon Oh , Tuovinen OH. Biodegradation of the phenoxy herbicides MCPP and 2,4-D in fixed-film column reactors. Int Biodeterior Biodegradation 1994; 33(1): 93-9.
[http://dx.doi.org/10.1016/0964-8305(94)90057-4]

[89]   Zhang TC, Fu YC, Bishop PL, *et al.* Transport and biodegradation of toxic organics in biofilms. J Hazard Mater 1995; 41(2-3): 267-85.
[http://dx.doi.org/10.1016/0304-3894(94)00118-Z]

[90]   Puhakka JA, Melin ES, Järvinen KT, *et al.* Fluidized-bed biofilms for chlorophenol mineralization. Water Sci Technol 1995; 31(1): 227-35.
[http://dx.doi.org/10.2166/wst.1995.0051]

[91]   Jin G, Englande AJ Jr. Carbon tetrachloride biodegradation in a fixed-biofilm reactor and its kinetic

study. Water Sci Technol 1998; 38(8-9): 155-62.
[http://dx.doi.org/10.2166/wst.1998.0802]

[92]   Yamaguchi T, Ishida M, Suzuki T. Biodegradation of hydrocarbons by *Prototheca zopfii* in rotating biological contactors. Process Biochem 1999; 35(3-4): 403-9.
[http://dx.doi.org/10.1016/S0032-9592(99)00086-2]

[93]   Eriksson M, Dalhammar G, Mohn WW. Bacterial growth and biofilm production on pyrene. FEMS Microbiol Ecol 2002; 40(1): 21-7.
[http://dx.doi.org/10.1111/j.1574-6941.2002.tb00932.x] [PMID: 19709207]

[94]   Kapdan IK, Kargi F. Simultaneous biodegradation and adsorption of textile dyestuff in an activated sludge unit. Process Biochem 2002; 37(9): 973-81.
[http://dx.doi.org/10.1016/S0032-9592(01)00309-0]

[95]   Vayenas DV, Michalopoulou E, Constantinides GN, Pavlou S, Payatakes AC. Visualization experiments of biodegradation in porous media and calculation of the biodegradation rate. Adv Water Resour 2002; 25(2): 203-19.
[http://dx.doi.org/10.1016/S0309-1708(01)00023-9]

[96]   Guiot SR, Tartakovsky B, Lanthier M, *et al.* Strategies for augmenting the pentachlorophenol degradation potential of UASB anaerobic granules. Water Sci Technol 2002; 45(10): 35-41.
[http://dx.doi.org/10.2166/wst.2002.0283] [PMID: 12188570]

[97]   Chang CC, Tseng SK, Chang CC, Ho CM. Degradation of 2-chlorophenol *via* a hydrogenotrophic biofilm under different reductive conditions. Chemosphere 2004; 56(10): 989-97.
[http://dx.doi.org/10.1016/j.chemosphere.2004.04.051] [PMID: 15268966]

[98]   Kargi F, Eker S. Removal of 2,4-dichlorophenol and toxicity from synthetic wastewater in a rotating perforated tube biofilm reactor. Process Biochem 2005; 40(6): 2105-11.
[http://dx.doi.org/10.1016/j.procbio.2004.07.013]

[99]   Dasgupta D, Ghosh R, Sengupta TK. Biofilm-mediated enhanced crude oil degradation by newly isolated *pseudomonas* species. ISRN Biotechnol 2013; 2013: 1-13.
[http://dx.doi.org/10.5402/2013/250749] [PMID: 25937972]

[100] Alessandrello MJ, Juárez Tomás MS, Raimondo EE, Vullo DL, Ferrero MA. Petroleum oil removal by immobilized bacterial cells on polyurethane foam under different temperature conditions. Mar Pollut Bull 2017; 122(1-2): 156-60.
[http://dx.doi.org/10.1016/j.marpolbul.2017.06.040] [PMID: 28641883]

[101] Lerch TZ, Chenu C, Dignac MF, Barriuso E, Mariotti A. Biofilm *vs.* Planktonic Lifestyle. consequences for Pesticide 2, 4-D metabolism by *Cupriavidusnecator* JMP134. Front Microbiol 2017; 8: 904.
[http://dx.doi.org/10.3389/fmicb.2017.00904] [PMID: 28588567]

[102] Parellada EA, Igarza M, Isacc P, *et al.* Squamocin, an annonaceous acetogenin, enhances naphthalene degradation mediated by *Bacillus atrophaeus* CN4. Rev Argent Microbiol 2017; 49(3): 282-8.
[http://dx.doi.org/10.1016/j.ram.2017.03.004] [PMID: 28554707]

[103] An X, Cheng Y, Huang M, *et al.* Treating organic cyanide-containing groundwater by immobilization of a nitrile-degrading bacterium with a biofilm-forming bacterium using fluidized bed reactors. Environ Pollut 2018; 237: 908-16.
[http://dx.doi.org/10.1016/j.envpol.2018.01.087] [PMID: 29551479]

[104] Kasai Y, Kishira H, Harayama S. Bacteria belonging to the genus *cycloclasticus* play a primary role in the degradation of aromatic hydrocarbons released in a marine environment. Appl Environ Microbiol 2002; 68(11): 5625-33.
[http://dx.doi.org/10.1128/AEM.68.11.5625-5633.2002] [PMID: 12406758]

[105] Srinath T, Verma T, Ramteke PW, Garg SK. Chromium (VI) biosorption and bioaccumulation by chromate resistant bacteria. Chemosphere 2002; 48(4): 427-35.

[http://dx.doi.org/10.1016/S0045-6535(02)00089-9] [PMID: 12152745]

[106]　Aguilar-Barajas E, Paluscio E, Cervantes C, Rensing C. Expression of chromate resistance genes from *Shewanella* sp. strain ANA-3 in *Escherichia coli*. FEMS Microbiol Lett 2008; 285(1): 97-100.
[http://dx.doi.org/10.1111/j.1574-6968.2008.01220.x] [PMID: 18537831]

[107]　Davis B, Ng SP, Palombo EA, Bhave M. A Tn 5051-like mer-containing transposon identified in a heavy metal tolerant strain *Achromobacter* sp. AO22. BMC Res Notes 2009; 2(1): 1-7.

[108]　Achal V, Kumari D, Pan X. Bioremediation of chromium contaminated soil by a brown-rot fungus, *Gloeophyllumsepiarium*. Research Journal of Microbiology 2011; 6(2): 166-71.
[http://dx.doi.org/10.3923/jm.2011.166.171]

[109]　Kumar Ramasamy R, Congeevaram S, Thamaraiselvi K. Evaluation of isolated fungal strain from e-waste recycling facility for effective sorption of toxic heavy metal Pb (II) ions and fungal protein molecular characterization-a mycoremediation approach. Asian J Exp Biol Sci 2011; 2: 342-7.

[110]　Achal V, Pan X, Fu Q, Zhang D. Biomineralization based remediation of As(III) contaminated soil by *Sporosarcina ginsengisoli*. J Hazard Mater 2012; 201-202: 178-84.
[http://dx.doi.org/10.1016/j.jhazmat.2011.11.067] [PMID: 22154871]

[111]　Özdemir S, Kilinc E, Poli A, Nicolaus B, Güven K. Cd, Cu, Ni, Mn and Zn resistance and bioaccumulation by thermophilic bacteria, *Geobacillus toebii* subsp. *decanicus* and *Geobacillus thermoleovorans* subsp. *stromboliensis*. World J Microbiol Biotechnol 2012; 28(1): 155-63.
[http://dx.doi.org/10.1007/s11274-011-0804-5] [PMID: 22806791]

[112]　Dash HR, Das S. Bioremediation of mercury and the importance of bacterial mer genes. Int Biodeterior Biodegradation 2012; 75: 207-13.
[http://dx.doi.org/10.1016/j.ibiod.2012.07.023]

[113]　Balamurugan D, Udayasooriyan C, Kamaladevi B. Chromium (VI) reduction by *Pseudomonas putida* and *Bacillus subtilis* isolated from contaminated soils. Int J Environ Sci 2014; 5(3): 522-9.

[114]　Banerjee S, Gothalwal R, Sahu PK, Sao S. Microbial observation in bioaccumulation of heavy metals from the ash dyke of thermal power plants of Chhattisgarh, India. Adv Biosci Biotechnol 2015; 6(2): 131-8.
[http://dx.doi.org/10.4236/abb.2015.62013]

CHAPTER 10

Microbial Processing for Valorization of Waste and Application

Muhammad Afzaal[1,*], **Farhan Saeed**[1], **Aftab Ahmad**[1], **Muhammad Saeed**[2], **Ifrah Usman**[1] and **Muhammad Nouman**[3]

[1] *Institute of Home & Food Sciences, Government College University, Faisalabad, Pakistan*

[2] *National Institute of Food Science & Technology, University of Agriculture, Faisalabad, Pakistan*

[3] *University Institute of Food Sciences & Technology, University of Lahore, Lahore, Pakistan*

Abstract: Most of the waste generated from agriculture and other industries is a great source of soil and water pollution. The increase in agriculture waste across the globe is of great concern because of various environmental and economic issues. However, genetic engineering and microbial processing development have helped extract various valuable products from this waste. Microbes have the natural potential to degrade this organic waste. This chapter highlights the opportunities to bio-valorize agricultural waste through microbes and produces valuable enzymes, biofuels and bioactive compounds. This chapter highlights how microbes may decrease the ever-increasing waste to produce various valuable products for industrial use.

Keywords: Agricultural waste, Economic, Microbes, Valuable Products.

INTRODUCTION

The global food demand is increasing continuously due to the ever-increasing global population, and that is thought to reach 9 billion people by the year 2050 [1]. The major sources of waste include agricultural wastes, industrial wastes and by-products, and discharge from commercial kitchen, household and hospitality sectors. Among the economical methods for waste, volarization is an ideal process for meeting the challenges of socio-economic concerns and environmental sustainability. Food waste can be described as loss generated in the last step of the food life cycle [2]. The generation of wastes from different sources also results in valuable resources. About 20-50% of the food waste consists of fruits and vegetable waste [3].

[*] **Corresponding author Muhammad Afzaal:** Institute of Home & Food Sciences, Government College University, Faisalabad, Pakistan, E-mail: muhammadafzaal@gcuf.edu.pk

Arun Karnwal & Abdel Rahman Mohammad Said Al-Tawaha (Eds.)

As a result of legume processing and utilization, various by-products are produced, which have social, economic, and nutritional importance [4]. The waste or by-products of legume crops include straw, leaves, pods, vines, sterns, *etc.* Sugarcane was traditionally cultivated to produce sugar, which currently fulfils 80% of sugar demand worldwide [5]. As a result of sugar production, large quantities of sugar industry waste or by-products such as bagasse, molasses, cane trash, press mud, *etc.*, are produced, which can be used to obtain valuable products. Agro-waste can be used for wastewater remediation, "diminishing waste by waste" [2]. Food waste disposal is achieved by composting and fermentation, and landfilling has been found more beneficial. The conventionally adopted incineration and dumping methods may result in environmental problems and health issues. Thus, using biological methods or microorganisms for waste management is an eco-friendly solution.

Sources of Food Wastes

Cereal and Pulses

Worldwide, a major portion of the human diet is based on cereals, which belong to the Gramineae family. These cereals include wheat, maize, rye, oat, *etc.* These cereals and pulses generate a large variety of wastes or by-products. A considerable quantity of by-products of pulses are generated in the subcontinent [6].

Vegetable and Fruits

Fruits and vegetables contain high-energy compounds such as polyphenol, carbohydrates, vitamins, valuable bioactive compounds, and fibres [7]. In addition, the moisture content of fruits and vegetables is also very high [8]. It is necessary to reduce the fruits and vegetable waste to improve the overall demand for food as well as the effectiveness of its supply chain.

Dairy

In Europe, approximately 29 million tons of wastage generate from dairy products. The reason for wastage is microbial attack and improper handling of dairy products. The waste from dairy includes organic matter, which aids in microbial growth; as a result, a diverse variety of products can be obtained [9, 10].

Meat and Poultry Industry

This industry covers a large part of the food chain. The wastes or by-products of meat include horns, bones, feathers, soft meat, skin, *etc.* [11]. The slaughterhouse products are being used for the fermentation process. As a result of fermentation, various types of products are produced having commercial significance. Similarly, it has been reported that many products (value-added) such as biomass could be obtained from slaughterhouse products or wastes [12].

Agro-Industrial Waste

The agro-wastes include organic substances such as shell, straw, leaves, bagasse *etc*. [13]. For solid-state fermentation and enzyme immobilization, agricultural waste or by-products are a very economical and useful source [14, 15]. Value-added products, for example, pigments, enzymes, antibiotics, phytochemicals and nutraceuticals, can be generated by feeding microorganisms on agro-industrial wastes.

Valorization and Agro-Waste Products

A huge range of products having commercial significance, for example, enzymes, biofuels, organic acids, biopolymers, dietary fibres, and nutraceuticals, have been made from bio-conversion of agro-waste (Fig. **1**) [16].

Valorization Technologies

Biological Treatments

1. **Composting** is a process which involves biological degradation of organic materials of heterogeneous nature under optimal conditions to obtain desired products or substances [17].

2. **An-aerobic digestion**, in this process, the conversion of organic matter (in the absence of oxygen) to the desired substance is carried out with the help of microorganisms. In this process, at various stages, decomposition is done by the action of bacteria [18].

Thermo-Chemical Treatment

1) Steam reforming is getting attention for producing hydrogen gas through natural gas. For hydrogen production by steam reforming, feedstocks such as methanol, jet fuel, petroleum gas and diesel are used [19].

2) Pyrolysis makes use of biomass for the generation of liquid fuel. An efficient choice of the thermal decomposition temperature and pyrolysis process can give a better yield for liquid fuel. The yield of bio-oil can be increased by approximately 80% at higher temperatures [20].

3) Torrefactionis an innovative thermal process which involves the conversion of wasted or cultured lignocellulosic feedstocks into a "charred" product. The product obtained by torrefaction may be utilized as fuel [21].

Useful Biological Transformation of Fruit Waste

Fruits waste can be transformed into a number of valuable products using different microbiological, physicochemical and enzymatic processes. The main processes used for transforming vegetable and fruit waste include the fermentation process (mostly Solid-State Fermentation), which also includes a pretreatment process in case of high lignin and cellulose-containing fruit waste.

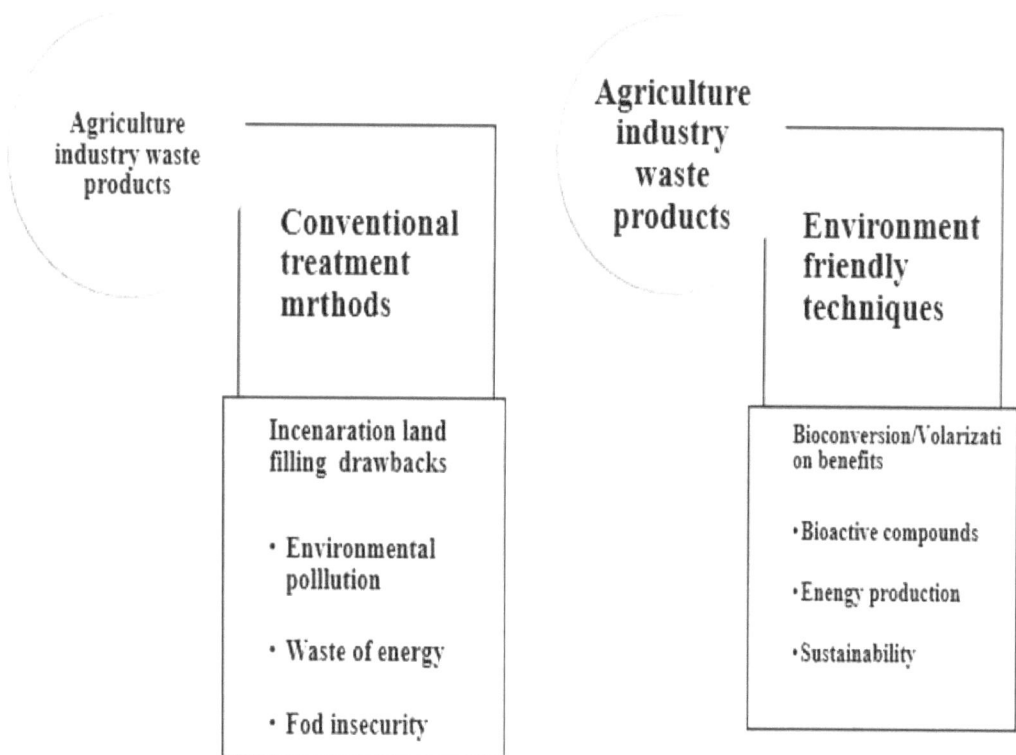

Fig. (1). Comparison of conventional treatment methods and eco-friendly treatment methods.

Pretreatment

This step is done before fermentation of fruit or vegetable waste in case of high lignin, cellulose or complex carbohydrate-containing fruits and vegetables to hydrolyze them into simpler fermentable products. It can be alkaline, acid or, in most cases, enzymatic hydrolysis. Research has shown that enzymatic hydrolysis yields the highest number of reduced sugars. These sugars are obtained *via* enzymatic tremulant industrial waste produced from fruit and vegetable processing [22].

Solid-State Fermentation (SSF)

Microbiological transformation of different types of waste, including fruits and vegetables, is carried by SSF. It is the fermentation process in which a moist solid substrate is used to grow microorganisms. Fruit waste is a favourable substrate for SSF because it is organic and includes several nutrients necessary for the growth of microorganisms, resulting in a higher yield.

SSF is an economical bio-processing that can be used to produce valuable products. This method is not only economical but also environmentally friendly [23]. As a result of this fermentation, a number of products like enzymes, alcohols, flavors, bioactive compounds, colourants, *etc.*, can be extracted depending on the type of substrate (fruit waste) used.

Different Valuable Products Obtained from the Biological Valorization of Fruit Waste

Following are some industrial products that can be obtained using microbiological processing of fruit waste.

Food Colourants

Natural food colours can be obtained by microbial fermentation of fruit waste. In a study, carotenoid was produced from apple pomace using Rhodotorula sp. ferment [24]. The yellow pigment was extracted from banana peel using *Monascuspurpureus* [25].

Natural Flavours and Fragrances

The biological treatment of fruit waste can produce many natural flavours. *Ceratocystisfimbriata* has been used to produce fruity aromatic compounds using apple pomace in fermentation [26]. Orange peel oil containing *D*-limonene has been used to transform *D*-limonene into a-terpineol, an aromatic compound [27].

Organic Acids

Production of organics acid is carried out by employing a fermentation process. So far, this is the most important acid widely produced by microbial bioprocessing. Citric acid has vast applications (acidity regulation, preservation and flavor enhancement.) in the food industry. The most important microorganism (*A. niger)* is utilized for the commercial formation of citric acid using various fruit wastes such as pineapple and grapes [28, 29]. Citric is mainly produced from apple pomace *via* fermentation [30]. Acetic acid and tartaric acid can also be produced using fruit waste.

Polysaccharides

Most of the commercial polysaccharides are produced from plants and algal sources. However, microbial polysaccharides have also been produced using SSF. Microbial polysaccharides are either in the cell wall or produced externally by the microorganism, called exopolysaccharides. Polysaccharides like xanthan gum, pectin and dietary fibres have been produced from fruit waste. Pectin can be obtained from mango and apple waste. Guava and lime waste, including peel and pulp, can be used to obtain dietary fibre to contain antioxidants and pectin [31, 32] Droxyalkanoates have been produced using cherry, apricot and grape waste by microbial fermentation by *Pseudomonas resinovorans* [33]

Edible Oil

Some microorganisms can produce oil called single cell oil up to 20% of their weight [34]. These oleaginous microorganisms include some yeasts, moulds, microalgae and bacteria. Some oleaginous microorganisms can produce oil that is comparable to oil produced from oilseeds. Some oleaginous yeasts like candida ssp. can produce oil by utilizing varying quantities of carbon. Therefore, fruit waste is a cheaper substrate for the production of oil from these. Single-cell oils are now used as a cheaper alternative to cocoa butter for a large-scale industrial production [35]

Enzymes

Fruits and vegetable waste has promising potential for producing industrial importance enzymes *via* SSF. Engineering of enzymes using food base waste is a very economical approach. *Aspergillus oryzae*is is a commonly used microorganism producing amylase enzyme from waste (apple and banana [36]. Different fruit wastes, including cranberry pomace, citrus peel and orange bagasse, are used to produce pectinase [37, 38]. All the enzymes involved in the pretreatment of waste (fruit and vegetable) can also be produced by SSF of fruit and vegetable waste.

Baker's Yeast

Baker's yeast has also been commercially produced using fruit waste. The production of Baker's yeast from fruit waste is a relatively cheaper process than using molasses as a substrate. Apple pomace has been used for this purpose, and it produces Baker's yeast with quality comparable to commercial Baker's yeast [39].

Useful Biological Transformation of Vegetable Waste

Vegetable waste can also be biologically processed using Solid-state fermentation and other enzymatic pretreatments like fruit waste to produce commercially valuable products. These products include biofuels, enzymes, flavours, single-cell proteins, organic acids, polysaccharides, single-cell oils, *etc.*

Flavours and Fragrances

The microbial processing of vegetable waste can produce some natural flavours and fragrances. Tomato and Pepper pomace have been used to produce tarhana and rose flavour, using fermentation by *K. marxianus*. *Kluyvero mycesmarxianus* and *Debaryo myceshansenii* have also been used for the fermentation of pepper pomace and tomato for the production of different esters and alcohols [39]. Some vegetable wastes like soybean and cassava waste can also be used to produce different flavours.

Organic Acids

As mentioned above, food-based waste has great potential for producing different types of organic acids. Similarly, fruit and vegetable wastes are being used for the same purpose around the world [40]. Lactic acid bacteria are extensively used in SSF for the said purpose. Carrot waste has good potential for citric acid production. Lactic acid bacteria are used for the process of fermentation. Sweet potato has also been used to produce oxalic acid [41].

Enzymes

The microbial processing of vegetable waste has also produced some enzymes. Potato peel has been used to produce amylase through solid-state fermentation by *Bacillusli cheniformis* and *Bacillus subtilis* [42].

Biofuels

Biofuels like biogas, different alcohols like biobutanol, bioethanol and bio hydrogen can be produced by the biological conversion of fruits and vegetable waste. Saccharification and fermentation are combined and used for the conversion of vegetable waste and fruit waste into biofuels. Fruits and vegetables contain starches and complex carbohydrates, which need to be broken down to simpler sugars for microorganisms to produce fermentation products, including alcohols and different gases depending upon the type of microorganisms and wastes used. For this purpose, enzymatic pretreatment is used to convert these complex carbohydrates like cellulose, lignin and starch to simpler fermentable sugars. A number of biofuels can be produced by this method. For example, *Clostridium pasteurianum* produces bio hydrogen [41], and different clostridium strains produce biobutanol [43, 44].

Polysaccharides

Some polysaccharides are also obtained commercially by the microbial processing of vegetable waste. Most bacteria are used for the industrial production of microbial polysaccharides. These bacteria belong to Xanthomonas, Leuconostoc, and Alcaligenes genera. Some polyhydroxy alkanoates can be produced by microbial processing of different vegetable wastes. For example, poly(R)-3-hydroxy butyrate can be produced by the microbial fermentation of saccharified potato starch using *Ralstoniaeutropha* NCIMB 11599 in conditions where Phosphate was limited [45]. Research has shown that some polyalkenoates can also be produced from residual cooking oil and lipids in fermentation by *Pseudomonas aeruginosa* [46].

Single-Cell Proteins (SCPs)

SCPs are obtained from microorganism culture, which are in the form of dried cells of microorganisms. SCPs contain 60-82% protein and are considered nutritionally good. Sweet potato bagasse and cassava are good substrates for producing single-cell proteins [47]. For the production of SCPs from starchy wastes, first liquefaction (to convert starch to dextrin) is done, followed by saccharification (to convert dextrin to fermentable sugars) and then fermentation. Potato beet, carrot and watermelon have also been used to produce single-cell proteins using different microorganisms [48, 49].

Useful Biological Transformation of Cereal Crops Waste

According to a study [50], by-products consist of damaged raw material leaves, straw, shells, seeds, bran, oil seed cakes, molasses, *etc*. (Fig **2**).

The three main cereal crops, rice (*Oryza sativa*), maize (*Zea mays*) and wheat (*Triticum sp*), give two-thirds of energy intake from food. Wheat, rice, maize, barley, sorghum, oat and millet are the top seven cereals. Cereal waste can be utilized as a source of bioactive compounds, *i.e.*, phenolic compounds, carotenoids, essential oil or beta-glucans. The extracted compounds from the agriculture by-products have a possible use as additives in different food industries because of having health benefits and antioxidant and antimicrobial properties; in addition to this, they can be utilized as a food colourant derived from natural sources. Besides, tissues from waste or by-products have flavour volatiles of varying concentrations, which can be recovered and utilized in different food industries. Additional valuable ingredients can be present in dietary fibres, including polysaccharides, proteins or lignin, which can be utilized as a functional food with a higher nutritional value. Similarly, several compounds in agricultural waste have antioxidant and antimicrobial effects, and are often utilized for edible coating or active packaging to enhance the shelf life of different food products. For instance, phenolic compounds such as hydroxyl tyrosyls in wheat bran or olive waste contain excellent antioxidant properties.

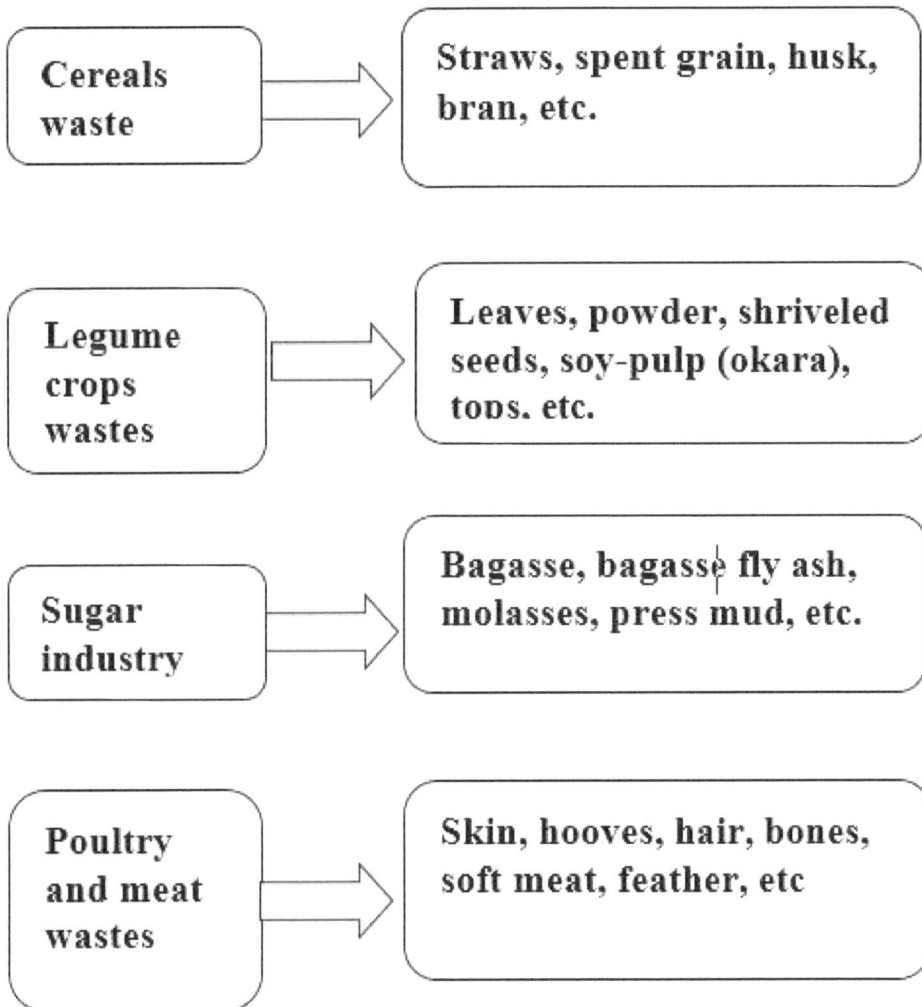

Cereals waste →	**Straws, spent grain, husk, bran, etc.**
Legume crops wastes →	**Leaves, powder, shriveled seeds, soy-pulp (okara), tops, etc.**
Sugar industry →	**Bagasse, bagasse fly ash, molasses, press mud, etc.**
Poultry and meat wastes →	**Skin, hooves, hair, bones, soft meat, feather, etc**

Fig. (2). Example of wastes from different sources.

Moreover, for packaging applications, residues of oat husk obtained from the processing of cereals can be utilized as a reinforcing agent in the production of bio-composites. Now the utilization of the waste stream includes the use of composting and biogas materials, plant fertilizers, and animal feed. Alternatively, they are landfilled or incinerated. Because of their composition, with a variety of valuable ingredients; waste or by-products can be valorized through several methods to achieving economic and environmental benefits.

Cereals and Their By-Products

By-products of wheat include wheat germ meal, fine bran, coarse bran, fine middling and coarse middling. Rice wastes are rice straw, rice husk and rice flakes . De-hulled rice is polished to give rice polishing, and broken rice and rice bran as by-products.

Similarly, by-products of corn include corn hull, corn husks, corn cob waste, corn stover waste, *etc.*, and sorghum includes sorghum bran, germ meal, gluten meal, and gluten feed. The beneficial compounds and products can be made by bioconversion of these wastes.

Production of Enzymes at an Industrial Scale

Agricultural waste has emerged as the best source for producing various valuable enzymes. That is why a range of food supply chain wastes have been utilized for this purpose. Oxidative enzymes such as laccase, cellulase, xylanase, amylase, lipase, and phytase have been the focus of production using agro-food waste. Microbial strains can degrade the complex polymer in plant biomass and utilize the sugar given for their nourishment. Due to this reason, the waste from the food processing industry can be used as raw material for the production of enzymes. In addition, high enzyme activity may be obtained by utilizing media optimization techniques. The method of solid stat fermentation is preferred over submerged fermentation, mostly owing to operation costs. The cost of operation of solid-state fermentation is $1/10^{th}$ to that of submerged fermentation. Moreover, solid-state fermentation causes the replication of the natural environment in a bioreactor, which has been shown to enhance enzyme yield [51].

Alpha-Amylase

This is a group of enzymes that smite alpha 1,4 linkages between glucose subunits in polysaccharides; as a result of enzymatic actions, these polysaccharides release subunit dextrin and oligomers. Many researchers have conducted various experimental studies for the optimization of alpha-amylase. They were using spent brewer's grain. The results indicated an overall increase (20%) in the yield.

The fungus *Aspergillus oryzae* was also used in another experiment with the optimized condition. Rajagopalan and Krishnan [52] used sugarcane bagasse as a substrate for *Bacillus subtilis* KCC103 for enzyme production and achieved a maximum alpha-amylase activity of 67.4 U mL-1.

Amyloglucosidase

Amyloglucosidase, also known as glucoamylases, can cleave the alpha-1,4 linkages and produce glucose subunits. These linkages are commonly present in starch. The research findings have shown wheat bran as a substrate in solid-state fermentation with *A. niger* giving sixfold enzyme *yield.* Numerous lignocellulose wastes are also used as a carbon source for producing amyl glucosidase. Additionally, it was reported that rice bran as a substrate is also useful for enzyme production. Ravindran [53] investigated that the production of fungal AMG agri-residues such as yam, cassava, and fruit peels has also been successfully utilized as an alternate carbon source.

Cellulase

These are very important enzymes, having great industrial importance. Cellulase plays a vital role in the production of ethanol on a commercial scale. These enzymes also have a key role in other allied industries for brewing, bread, paper and pulp, and detergent [54].

Xylanase

These enzymes have a significant role in the breakdown of xylan. This polysaccharide (Xylan) consists of xylose-residue. Each unit of this polysaccharide is linked by a beta-1,4-glycosidic bond. It was investigated that wheat straw as a substrate has a key role in the production of xylanase.

Inulinase

Fructose and fructose-oligosaccharides are the most favourite consumed sweeteners. Many researchers investigated the production of inulinase by utilizing lignocellulosic substrate. Wheat, rice and sugar bagasse are suitable substrates with *saccharomyces sp.* for the production of inulinase.

Beta-Glucanase

In the endosperm cell wall of cereals for example oats, sorghum, rye, and barley1,3-1,4-beta glucans are found. Using lignocellulosic industry waste as a substrate, the various microorganisms have been tested for the production of beta-glucanase. Oatmeal is a suitable carbon source under solid-state fermentation conditions.

Invertase

It is a glycoprotein that can hydrolyze sucrose to dextrose and fructose, also known as beta-fructo furanosidase (EC.3.2.1.26). Optimum conditions for invertase activity are a temperature of 55 degrees C and pH 4.5. For the production of invertase enzyme in industry, the main microbe is *Saccharomyces cerevisiae* for the production of invertase enzyme employing different microorganisms, and wheat bran bagasse waste proved to be excellent carbon source substitutes [55].

Table 1. Utilization of agro-waste for valuable products generation.

Agro Waste Products	Enzyme	Micro Organism	Value Added Products Obtained	Source
Wheat straw and Pearl millet	Laccase	(Fungus) *Pleurusotostreatus*	Bio-methane and reducing sugar production observed	[56]
Moringa tree stalks		*Trichoderma reesei*coupled with *Saccharomyces cerevisiae*	animal(lamb) feed with improved nutritive value observed	
Rice straw	Cellulose, laccase and xylanase	*Coprinopsiscinerea*	Improved Enzyme's activity profile observed	[57]
Sugar-cane bagasse and rice straw	Cellulase	*Aspergillus tubingensis*	Production of Fermentable sugar observed	[57]
Millet straw and wheat straw	Lacasse	(Fungus) *Chaetomium globosporum*	Bio gas and reducing sugar production observed	[56]
Wheat straw		(Yeast) *Candida guilliermondii*	Xylitol	[56]
Hay	Lignocellulolytic enzymes	(Fungus) *Neocallimastix frontalis*	Biogas production observed	[57]

Bioconversion Process of Food Waste for the Production of Value-added Products (Table 1)

Organic Acids

In the food, textile and pharmaceutical industries, organic acids are used for different purposes. A group of these organic acids can be formed using bacterial fermentation. The Lactobacillus group is very common for producing lactic acid using agro-industrial waste. The high acid yield can be obtained through proper time-temperature combination in fermentation. Corn steep liquor (CSR) was utilized as a single and inexpensive source of nitrogen, and some components are uses as an alternative to yeast extract for a cost-effective and valuable yield of lactic acid. Various agro-wastes are being used across the world for the production of various acids. However, rice bran is one of the most important wastes. Rice bran is a major carbon source, and *L. rhamnosusis* is considered the most suitable fermenting organism. Amongst the world's microbial processes used in industry, fermentation of citric acid by using *Aspergillus niger* is the most significant. The same fungus was utilized to evaluate economic citric acid production by solid-state fermentation, using untreated brewer's spent grain agro-waste. Two wheat by-products were pretreated enzymatically and were utilized to obtain succinic acid by *Actinobacillus succinogenes.* In solid-state fermentation, wheat bran was used as a substrate for *Aspergillus* strains to manufacture glucoamylase and protease enzymes, which are consumed to pretreat wheat gluten or wheat bran. It was observed that the succinic acid production was 22g/L when wheat gluten was used while it was 50.6g/L with wheat bran substrate. Furthermore, for succinic acid production by acid pretreated corn fibre by-products, A. *succinogenes* was used aerobically, and as a result, 35.5 g/L was obtained [58].

Flavouring Agent (Vanillin)

Along with numerous uses in the food industry, vanillin is used as a flavouring agent and, on account of its antimicrobial and antioxidant properties, it is also used as a preservative. Moreover, it also has application in herbicides and pharmaceuticals. Ferulic acid is a vanillin's precursor, which can be obtained from agro-waste by enzymatic treatments. Ferulic acid is obtained by bioprocess using the *E. coli* strain. This ferulic acid is a very important substrate for the production of vanillin. It was reported that corn cob on hydrolysis gave 171mg/L ferulic acid that later on was utilized as a medium for the vanillin [59].

Nutraceuticals

Many researchers described that value-added chemicals, *i.e.* dietary fibres and antioxidants, can be extracted by bio-conversion of cereals waste. As a result of rice milling, rice bran is produced, which is rich in minerals, proteins, and phytochemicals revealed that the antioxidant activity of bread increased fivefold by the addition of rice bran to the wheat flour by 30% [60].

Antibiotics

Different microorganisms produce these substances and can inhibit growth or kill other microbes. Different agriculture wastes or by-products can be used to produce antibiotics. For the production of antibiotics, cereals waste such as rice hulls, corn cobs, and saw dust was used., *Streptomyces rimosus* successfully produced oxytetracycline with the solid-state fermentation method using a groundnut shell as a substrate [61].

Useful Biological Transformation of Legume Crops

Various legume crops of the legume family are found all over the world. Except for major legumes, statistical data on the majority of legumes is unavailable. Grain fractions of various legume crops showed that the husk or seed coat of legume contains high fibre content (20-32%) and protein content. The nutritional importance of legumes by-products as a ruminant feed source is noticeable. In India, the by-products of chickpea milling are fed to cattle. Additional feed and food sources can be obtained from various legumes. The roots of cowpeas and sweet potatoes are eaten in East Africa. After cereals, the second major source of food is legumes by-products due to their high protein content. Green parts of pigeon pea and black gram provide fodder, and dried vines of chickpea can be used as cattle feed; in addition to this, broad cowpea beans can be utilized for hay and silage. Milling pulses on commercial scale can yield nearly 25% by products consisting of husk 14%, powder 12%, and broken or shrivelled seeds 13%. Mostly legume processing is performed in a limited speciality and small-scale facilities; therefore, the recovery of proteins and other valuable nutrients from legume by-products is limited. In the production of soya milk and tofu, as a result of soya bean processing, a by-product called Okara is obtained, which contains approximately 25-33% protein. Annually large quantity of Okara is produced, and proper disposal or reuse of such material is necessary [62, 63]

Useful Products Generation by the Valorization of Legume by-Products

Oncom Production

Oncom is a very famous Indonesian fermented product produced using various agro-based waste products. There are mainly three types of Oncom. The most famous oncom type can be obtained from a peanut oil by-product known as peanut press-cake. This product (oncomkacang) is very famous in West Java. The second type (Oncomtahoo) popular in Jakarta is obtained from soya waste. The third type (oncomampashunkwe) is produced from the waste of starch flour. However, the safety issues of the traditional fermented products still need to be addressed [64, 65].

Tempeh Production

This is a fermented product and is popular around the world in developing as well as developed countries. In Malaysia and Indonesia, tempeh is prepared in small industries or on a small scale. Its aroma and texture are better than non-fermented products. Instead of wasting soya bean milk, waste can be utilized to make a product having high protein content. Tempeh is available in different forms. Soft products, *i.e.* Tempeh, can be prepared by utilization of boiled soya bean. For the production of Tempeh, *Rhizopus* strains, due to having abilities to degrade raw material, are used. Soya bean milk waste can be used to produce Tempeh with superior nutritional quality, such as high protein content and also proved cost-effective [66, 67].

Mushroom *Production*

Mushrooms can be hypogynous and epigenous, consisting of a unique fruiting body. Mushroom has superior medicinal properties and can be utilized as protein-rich food. By using agro-waste, they have cultivated Pleurotus mushrooms. Their study observed improved carbohydrates, lipids, and proteins using agricultural waste straw as a substrate. Therefore, they recommended using edible oyster mushrooms as a high protein content food. Paddy straws (as a substrate) are very important for the best production of various types of mushrooms [68, 69].

Production of Antibiotic

Oil pressed cake (as substrate) is most useful for producing antibiotics (rifamycin B) through solid-stat fermentation. Additionally, coconut oil cake and groundnut shell gave maximum antibiotic production [70].

Useful Biological Transformation of Sugar Industry Waste

While harvesting one tone of sugar cane, approximately 270 kg of sugar cane leaves are obtained as waste material. By processing one tone sugarcane, 0.3 tons of bagasse (wet basis), 0.03 tons of press mud, and 0.041tons of molasses were generated [71].

Bio Conversion of the Sugar Industry by-Products Molasses

Molasses is frequently used in fermentation as a substrate to generate ethanol, resulting in "spent wash, " which is an aqueous waste. Contrary to this, in the broiler, inefficient burning of bagasse results in heat generation.

Bagasse

After trituration and extraction of juice from sugar cane, bagasse is produced. It is considered the most significant by-product of the sugarcane industry. In the sugar industry, using the leftover bagasse to produce ethanol helped to improve air quality compared to open-burning of bagasse. In various studies for inulinase production, yacon and sugar cane bagasse have been utilized as substrates. Application of genetically modified strain of *L. delbrueckii* subsp. *delbrueckii* mutant Uc-3 succeeded in generating 67g/L of metabolite with a yield of 80%. There are 50% w/w sugars in molasses; sucrose is the most abundant. A viscous liquid is produced due to high sugar concentration, which may cause substrate inhibition. In addition, metals such as iron, zinc, copper, *etc.*, present in molasses may cause the inactivation of useful enzymes and can also affect microbial growth. A benefit of this by-product is its suitable buffer capacity, which proves to be useful during fermentation of lactic acid by maintaining a pH higher than five, and there will be no need to add neutralizers to the culture medium. In most of the studies, it was reported that by utilizing molasses for lactic acid production, results were observed between 40 and 166 g/L, And the volumetric productivity ranges between 2.00 to 4.20 g/L with a yield range between 0.80 and 0.98 g/g [72 - 75].

Press Mud

During the sugar cane juice clarification step, an amorphous residue is obtained having dark brownish color called press mud. According to a study [71], it is produced at 0.03 tone per ton of sugarcane processing. For sugar purification, the carbonation and sulphuration process affect the press mud composition. In China and Brazil, press mud in combination with bagasse ash or spent wash is used as a fertilizer in land application practices. The press mud decomposition is harmful to the environment due to greenhouse gases emission and leaching, which can cause pollution [76, 77].

Spent Wash

The deep brown wastewater produced in the ethanol distilleries during distillation is called spent wash. It is also known as vinasse. It shows a pH range of 3.5-5 which means it is acidic. During per litre ethanol production, 10-15 litre of spent wash is generated. One of the eco-friendly applications of valorization is the conversion of the by-product of sugarcane, *i.e.* leaves, into biochar which is useful for cooking applications. By effectively managing sugar industry wastes such as bagasse ash and spent wash, the aim of reducing the environmental burden due to fertilizers and pesticides can be achieved. In addition, higher yields can be obtained by utilizing desired quantities of these wastes. All this highlights the significance of valorization for achieving sustainability and food security [78].

Utilization of Food Waste for Energy Generation

Biodiesel

Biodiesel has come forward as an alternative fuel, which fats and oils can obtain. Its production is costly due to expensive feedstock. Thus, with the aim of cost reduction, food waste can be used as inexpensive feedstock. Utilizing the transesterification process, biodiesel can be generated from food waste, acid/alkali catalyst, or microbial oils generated by oleaginous microorganisms [79, 80].

Biohydrogen

Food waste rich in carbohydrates has twenty times more potential for generating bio hydrogen than food rich in protein and fat. Bio hydrogen is an efficient source of renewable energy, with a yield of 142 MJ/kg energy. Food waste has been utilized for the production of biohydrogen using batch fermentation, single fermentation, and continuous as well as semi-continuous processes [80]

Electric Power Generation

Mostly conventional methods such as landfill, incineration or burning were used for food disposal. These methods can cause groundwater contamination in addition to the generation of toxic gases and health-hazardous compounds. Therefore, food waste utilization for energy production is a pollution reduction and sustainable approach. An aerobically treating food waste in precise conditions and devices, for example, microbial fuel cell electric power can be generated. Thus microorganisms can be effectively utilized for biotransformation of wastes or by-products, and valorization for food waste management has proven to be an eco-friendly and sustainable approach [81 - 83].

CONCLUSION

Throughout the world, food is one of the major sectors contributing to waste generation. The huge and ever-growing amount of agro-waste turns out to be a major concern. Improper management leads to environmental pollution and a threat to sustainability. Volarization is a promising approach to maintaining a sustainable food system. As a result of volarization by the action of microbes, products such as biofuel, biochemicals, biomolecules, electrical energy and other value-added products can be generated.

CONSENT FOR PUBLICATION

Not applicable.

CONFLICT OF INTEREST

The author declares no conflict of interest, financial or otherwise.

ACKNOWLEDGEMENT

Declared none.

REFERENCES

[1]   Wong JWC, Kaur G, Mehariya S, Karthikeyan OP, Chen G. Food waste treatment by anaerobic co-digestion with saline sludge and its implications for energy recovery in Hong Kong. Bioresour Technol 2018; 268: 824-8.
[http://dx.doi.org/10.1016/j.biortech.2018.07.113] [PMID: 30064901]

[2]   Xiong X, Yu IKM, Tsang DCW, *et al.* Value-added chemicals from food supply chain wastes: State-of-the-art review and future prospects. Chem Eng J 2019; 375: 121983.
[http://dx.doi.org/10.1016/j.cej.2019.121983]

[3]   Abdelradi F. Food waste behaviour at the household level: A conceptual framework. Waste Manag 2018; 1(71): 485-93.
[http://dx.doi.org/10.1016/j.wasman.2017.10.001]

[4]   Baiano A. Recovery of biomolecules from food wastes-a review. Molecules 2014; 19(9): 14821-42.
[http://dx.doi.org/10.3390/molecules190914821] [PMID: 25232705]

[5]   Dotaniya ML, Datta SC, Biswas DR, *et al.* Use of sugarcane industrial by-products for improving sugarcane productivity and soil health. Int J Recycl Org Waste Agric 2016; 5(3): 185-94.
[http://dx.doi.org/10.1007/s40093-016-0132-8]

[6]   Medhe S, Anand M, Anal AK. Dietary fibers, dietary peptides and dietary essential fatty acids from food processing by-products. Food processing by-products and their utilization 2017; 6: 111-36.
[http://dx.doi.org/10.1002/9781118432921.ch6]

[7]   Schieber A. Side streams of plant food processing as a source of valuable compounds: Selected examples. Annu Rev Food Sci Technol 2017; 8(1): 97-112.
[http://dx.doi.org/10.1146/annurev-food-030216-030135] [PMID: 28068488]

[8]   Matharu AS, de Melo EM, Houghton JA. Opportunity for high value-added chemicals from food supply chain wastes. Bioresour Technol 2016; 215: 123-30.
[http://dx.doi.org/10.1016/j.biortech.2016.03.039] [PMID: 26996261]

[9]   Mahboubi A, Ferreira J, Taherzadeh M, Lennartsson P. Production of fungal biomass for feed, fatty acids, and glycerol by Aspergillus oryzae from fat-rich dairy substrates. Fermentation (Basel) 2017; 3(4): 48.
[http://dx.doi.org/10.3390/fermentation3040048]

[10]   Lappa I, Papadaki A, Kachrimanidou V, *et al.* Cheese whey processing: integrated biorefinery concepts and emerging food applications. Foods 2019; 8(8): 347.
[http://dx.doi.org/10.3390/foods8080347] [PMID: 31443236]

[11]   Ashayerizadeh O, Dastar B, Samadi F, Khomeiri M, Yamchi A, Zerehdaran S. Study on the chemical and microbial composition and probiotic characteristics of dominant lactic acid bacteria in fermented poultry slaughterhouse waste. Waste Manag 2017; 65: 178-85.
[http://dx.doi.org/10.1016/j.wasman.2017.04.017] [PMID: 28408278]

[12]   Shaheen MS, Mehmood S, Mahmud A, *et al.* Effect of Different Brooding Sources on Growth, Blood Glucose, Cholesterol and Economic Appraisal of Three Commercial Broiler Strains. Pak J Zool 2019; 51(2)
[http://dx.doi.org/10.17582/journal.pjz/2019.51.2.575.582]

[13] Sadh PK, Duhan S, Duhan JS. Agro-industrial wastes and their utilization using solid state fermentation: a review. Bioresour Bioprocess 2018; 5(1): 1-5.
[http://dx.doi.org/10.1186/s40643-017-0187-z]

[14] Bogar B, Szakacs G, Linden JC, Pandey A, Tengerdy RP. Optimization of phytase production by solid substrate fermentation. J Ind Microbiol Biotechnol 2003; 30(3): 183-9.
[http://dx.doi.org/10.1007/s10295-003-0027-3] [PMID: 12715256]

[15] Singh KK, Talat M, Hasan SH. Removal of lead from aqueous solutions by agricultural waste maize bran. Bioresour Technol 2006; 97(16): 2124-30.
[http://dx.doi.org/10.1016/j.biortech.2005.09.016] [PMID: 16275062]

[16] Galanakis CM. Recovery of high added-value components from food wastes: Conventional, emerging technologies and commercialized applications. Trends Food Sci Technol 2012; 26(2): 68-87.
[http://dx.doi.org/10.1016/j.tifs.2012.03.003]

[17] 2016.

[18] Ward AJ, Hobbs PJ, Holliman PJ, Jones DL. Optimisation of the anaerobic digestion of agricultural resources. Bioresour Technol 2008; 99(17): 7928-40.
[http://dx.doi.org/10.1016/j.biortech.2008.02.044] [PMID: 18406612]

[19] Singh Siwal S, Zhang Q, Sun C, Thakur S, Kumar Gupta V, Kumar Thakur V. Energy production from steam gasification processes and parameters that contemplate in biomass gasifier – A review. Bioresour Technol 2020; 297: 122481.
[http://dx.doi.org/10.1016/j.biortech.2019.122481] [PMID: 31796379]

[20] Antal MJ. Biomass pyrolysis: A review of the literature Part 2—lignocellulose pyrolysis. Advances in solar energy 1985; 175-255.

[21] Bergman PC. a R. Boersma, RWR Zwart, and JH a Kiel, "Torrefaction for biomass co-firing in existing coal-fired power stations,". Energy Res Cent Netherlands ECN 2005; ECNC05013: 71.

[22] Patle S, Lal B. Ethanol production from hydrolysed agricultural wastes using mixed culture of Zymomonas mobilis and Candida tropicalis. Biotechnol Lett 2007; 29(12): 1839-43.
[http://dx.doi.org/10.1007/s10529-007-9493-4] [PMID: 17657407]

[23] Ajila CM, Brar SK, Verma M, Tyagi RD, Godbout S, Valéro JR. Bio-processing of agro-byproducts to animal feed. Crit Rev Biotechnol 2012; 32(4): 382-400.
[http://dx.doi.org/10.3109/07388551.2012.659172] [PMID: 22380921]

[24] Joshi VK, Attri D, Bala A, Bhushan S. Microbial Pigments. Indian J Biotech 2003; 2: 362-69.

[25] Panda SK, Ray RC, Mishra SS, Kayitesi E. Microbial processing of fruit and vegetable wastes into potential biocommodities: a review. Crit Rev Biotechnol 2018; 38(1): 1-16.
[http://dx.doi.org/10.1080/07388551.2017.1311295] [PMID: 28462596]

[26] Adi DD, Oduro I, Simpson BK. Biological and microbial technologies for the transformation of fruits and vegetable wastes. Byproducts from Agriculture and Fisheries: Adding Value for Food, Feed, Pharma, and Fuels 2019; 403-20.
[http://dx.doi.org/10.1002/9781119383956.ch17]

[27] Badee AZM, Helmy SA, Morsy NFS. Utilisation of orange peel in the production of α-terpineol by Penicillium digitatum (NRRL 1202). Food Chem 2011; 126(3): 849-54.
[http://dx.doi.org/10.1016/j.foodchem.2010.11.046]

[28] Yalcin SK, Bozdemir MT, Ozbas ZY. Citric acid production by yeasts: fermentation conditions, process optimization and strain improvement. Current research, technology and education topics in applied microbiology and microbial biotechnology 2010; 9: 1374-82.

[29] Kumar D, Jain VK, Shanker G, Srivastava A. Utilisation of fruits waste for citric acid production by solid state fermentation. Process Biochem 2003; 38(12): 1725-9.
[http://dx.doi.org/10.1016/S0032-9592(02)00253-4]

[30]   Bhushan S, Kalia K, Sharma M, Singh B, Ahuja PS. Processing of apple pomace for bioactive molecules. Crit Rev Biotechnol 2008; 28(4): 285-96.
[http://dx.doi.org/10.1080/07388550802368895] [PMID: 19051107]

[31]   Esparza I, Jiménez-Moreno N, Bimbela F, Ancín-Azpilicueta C, Gandía LM. Fruit and vegetable waste management: Conventional and emerging approaches. J Environ Manage 2020; 265: 110510.
[http://dx.doi.org/10.1016/j.jenvman.2020.110510] [PMID: 32275240]

[32]   Schieber A, Hilt P, Streker P, Endreß HU, Rentschler C, Carle R. A new process for the combined recovery of pectin and phenolic compounds from apple pomace. Innov Food Sci Emerg Technol 2003; 4(1): 99-107.
[http://dx.doi.org/10.1016/S1466-8564(02)00087-5]

[33]   Follonier S, Goyder MS, Silvestri AC, *et al.* Fruit pomace and waste frying oil as sustainable resources for the bioproduction of medium-chain-length polyhydroxyalkanoates. Int J Biol Macromol 2014; 71: 42-52.
[http://dx.doi.org/10.1016/j.ijbiomac.2014.05.061] [PMID: 24882726]

[34]   Thevenieau F, Nicaud JM. Microorganisms as sources of oils. OCL 2013; 20(6): D603.
[http://dx.doi.org/10.1051/ocl/2013034]

[35]   Papanikolaou S, Aggelis G. Lipids of oleaginous yeasts. Part II: Technology and potential applications. Eur J Lipid Sci Technol 2011; 113(8): 1052-73.
[http://dx.doi.org/10.1002/ejlt.201100015]

[36]   Nigam P, Singh D. Solid-state (substrate) fermentation systems and their applications in biotechnology. J Basic Microbiol 1994; 34(6): 405-23.
[http://dx.doi.org/10.1002/jobm.3620340607]

[37]   Dhillon SS, Gill RK, Gill SS, Singh M. Studies on the utilization of citrus peel for pectinase production using fungus *Aspergillus niger*. Int J Environ Stud 2004; 61(2): 199-210.
[http://dx.doi.org/10.1080/0020723032000143346]

[38]   Zheng Z, Shetty K. Solid state production of polygalacturonase by Lentinus edodes using fruit processing wastes. Process Biochem 2000; 35(8): 825-30.
[http://dx.doi.org/10.1016/S0032-9592(99)00143-0]

[39]   Güneşer O, Demirkol A, Karagül Yüceer Y, Özmen Toğay S, İşleten Hoşoğlu M, Elibol M. Bioflavour production from tomato and pepper pomaces by Kluyveromyces marxianus and Debaryomyces hansenii. Bioprocess Biosyst Eng 2015; 38(6): 1143-55.
[http://dx.doi.org/10.1007/s00449-015-1356-0] [PMID: 25614449]

[40]   Soccol CR, Vandenberghe LP, Rodrigues C, Medeiros AB, Larroche C, Pandey A. Production of organic acids by solid-state fermentation. InCurrent Developments in Solid-State Fermentation 2008; 205-29.
[http://dx.doi.org/10.1007/978-0-387-75213-6_10]

[41]   Ajila CM, Brar SK, Verma M, Rao UP. Sustainable solutions for agro processing waste management: an overview. Environmental protection strategies for sustainable development. 2012; 65-109.
[http://dx.doi.org/10.1007/978-94-007-1591-2_3]

[42]   Shukla J, Kar R. Potato peel as a solid state substrate for thermostable α-amylase production by thermophilic Bacillus isolates. World J Microbiol Biotechnol 2006; 22(5): 417-22.
[http://dx.doi.org/10.1007/s11274-005-9049-5]

[43]   Khedkar MA, Nimbalkar PR, Gaikwad SG, Chavan PV, Bankar SB. Sustainable biobutanol production from pineapple waste by using *Clostridium acetobutylicum* B 527: Drying kinetics study. Bioresour Technol 2017; 225: 359-66.
[http://dx.doi.org/10.1016/j.biortech.2016.11.058] [PMID: 27939964]

[44]   Nimbalkar PR, Khedkar MA, Chavan PV, Bankar SB. Biobutanol production using pea pod waste as substrate: Impact of drying on saccharification and fermentation. Renew Energy 2018; 117: 520-9.

[http://dx.doi.org/10.1016/j.renene.2017.10.079]

[45]   Haas R, Jin B, Zepf FT. Production of Poly(3-hydroxybutyrate) from waste potato starch. Biosci Biotechnol Biochem 2008; 72(1): 253-6.
[http://dx.doi.org/10.1271/bbb.70503] [PMID: 18175895]

[46]   Fernández D, Rodríguez E, Bassas M, *et al.* Agro-industrial oily wastes as substrates for PHA production by the new strain *Pseudomonas aeruginosa* NCIB 40045: Effect of culture conditions. Biochem Eng J 2005; 26(2-3): 159-67.
[http://dx.doi.org/10.1016/j.bej.2005.04.022]

[47]   Ward OP, Singh A, Ray RC. Single Cell Protein from Horticultural and Related Food Processing Wastes. Microbial Biotechnology in Horticulture 2008; 3(3): 273.

[48]   Oliveira MA, Reis EM, Nozaki J. Biological treatment of wastewater from the cassava meal industry. Environ Res 2001; 85(2): 177-83.
[http://dx.doi.org/10.1006/enrs.2000.4118] [PMID: 11161666]

[49]   Stabnikova O, Wang JY, Ding HB, Tay JH. Biotransformation of vegetable and fruit processing wastes into yeast biomass enriched with selenium. Bioresour Technol 2005; 96(6): 747-51.
[http://dx.doi.org/10.1016/j.biortech.2004.06.022] [PMID: 15712407]

[50]   Sharma P, Gaur VK, Kim SH, Pandey A. Microbial strategies for bio-transforming food waste into resources. Bioresour Technol 2020; 299: 122580.
[http://dx.doi.org/10.1016/j.biortech.2019.122580] [PMID: 31877479]

[51]   Barrios-González J. Solid-state fermentation: Physiology of solid medium, its molecular basis and applications. Process Biochem 2012; 47(2): 175-85.
[http://dx.doi.org/10.1016/j.procbio.2011.11.016]

[52]   Rajagopalan G, Krishnan C. α-Amylase production from catabolite derepressed *Bacillus subtilis* KCC103 utilizing sugarcane bagasse hydrolysate. Bioresour Technol 2008; 99(8): 3044-50.
[http://dx.doi.org/10.1016/j.biortech.2007.06.001] [PMID: 17644331]

[53]   Ravindran R, Hassan S, Williams G, Jaiswal A. A review on bioconversion of agro-industrial wastes to industrially important enzymes. Bioengineering (Basel) 2018; 5(4): 93.
[http://dx.doi.org/10.3390/bioengineering5040093] [PMID: 30373279]

[54]   Krishna C. Solid-state fermentation systems-an overview. Crit Rev Biotechnol 2005; 25(1-2): 1-30.
[http://dx.doi.org/10.1080/07388550590925383] [PMID: 15999850]

[55]   Ohara A, Soares de Castro RJ, Goia Nishide T, Gonçalves Dias FF, Pavan Bagagli M, Harumi Sato H. Invertase production by Aspergillus niger under solid state fermentation: Focus on physical–chemical parameters, synergistic and antagonistic effects using agro-industrial wastes. Biocatal Agric Biotechnol 2015; 4(4): 645-52.
[http://dx.doi.org/10.1016/j.bcab.2015.06.008]

[56]   Yadav RL, Solomon S. Potential of developing sugarcane by-product based industries in India. Sugar Tech 2006; 8(2-3): 104-11.
[http://dx.doi.org/10.1007/BF02943642]

[57]   Chen T, Luo L, Deng S, *et al.* Sorption of tetracycline on H_3PO_4 modified biochar derived from rice straw and swine manure. Bioresour Technol 2018; 267: 431-7.
[http://dx.doi.org/10.1016/j.biortech.2018.07.074] [PMID: 30032057]

[58]   Saha BC, Cotta MA. Enzymatic hydrolysis and fermentation of lime pretreated wheat straw to ethanol. Journal of Chemical Technology & Biotechnology: International Research in Process. Environmental & Clean Technology 2007; 82(10): 913-9.

[59]   Banerjee G, Chattopadhyay P. Vanillin biotechnology: the perspectives and future. J Sci Food Agric 2019; 99(2): 499-506.
[http://dx.doi.org/10.1002/jsfa.9303] [PMID: 30094833]

[60] Gul K, Yousuf B, Singh AK, Singh P, Wani AA. Rice bran: Nutritional values and its emerging potential for development of functional food—A review. Bioactive Carbohydrates and Dietary Fibre 2015; 6(1): 24-30.
[http://dx.doi.org/10.1016/j.bcdf.2015.06.002]

[61] Chen Z, Li Y, Peng Y, Ye C, Zhang S. Effects of antibiotics on hydrolase activity and structure of microbial community during aerobic co-composting of food waste with sewage sludge. Bioresour Technol 2021; 321: 124506.
[http://dx.doi.org/10.1016/j.biortech.2020.124506] [PMID: 33310386]

[62] Nielsen C, Rahman A, Rehman AU, Walsh MK, Miller CD. Food waste conversion to microbial polyhydroxyalkanoates. Microb Biotechnol 2017; 10(6): 1338-52.
[http://dx.doi.org/10.1111/1751-7915.12776] [PMID: 28736901]

[63] Mustafa R, Reaney MJ. Aquafaba, from Food Waste to a Value Added Product. Food Wastes and By-products: Nutraceutical and Health Potential. 2020.
[http://dx.doi.org/10.1002/9781119534167.ch4]

[64] Reddy NR, Pierson MD, Sathe SK, Salunkhe DK, Beuchat LR. Legume based fermented foods: Their preparation and nutritional quality. CRC Crit Rev Food Sci Nutr 1983; 17(4): 335-70.
[http://dx.doi.org/10.1080/10408398209527353] [PMID: 6759047]

[65] Sharma A. A review on traditional technology and safety challenges with regard to antinutrients in legume foods. J Food Sci Technol 2020; 1-21.
[PMID: 34294949]

[66] Nakajima N, Nozaki N, Ishihara K, Ishikawa A, Tsuji H. Analysis of isoflavone content in tempeh, a fermented soybean, and preparation of a new isoflavone-enriched tempeh. J Biosci Bioeng 2005; 100(6): 685-7.
[http://dx.doi.org/10.1263/jbb.100.685] [PMID: 16473782]

[67] Bavia ACF, Silva CE, Ferreira MP, Leite RS, Mandarino JMG, Carrão-Panizzi MC. Chemical composition of tempeh from soybean cultivars specially developed for human consumption. Food Sci Technol (Campinas) 2012; 32(3): 613-20.
[http://dx.doi.org/10.1590/S0101-20612012005000085]

[68] Srianta I, Kusdiyantini E, Zubaidah E, *et al.* Utilization of agro-industrial by-products in Monascus fermentation: a review. Bioresour Bioprocess 2021; 8(1): 129.
[http://dx.doi.org/10.1186/s40643-021-00473-4]

[69] Wiafe-Kwagyan M.

[70] Haq A, Siddiqi M, Batool SZ, *et al.* Comprehensive investigation on the synergistic antibacterial activities of Jatropha curcas pressed cake and seed oil in combination with antibiotics. AMB Express 2019; 9(1): 67.
[http://dx.doi.org/10.1186/s13568-019-0793-6] [PMID: 31102037]

[71] Gupta N, Tripathi S, Balomajumder C. Characterization of pressmud: A sugar industry waste. Fuel 2011; 90(1): 389-94.
[http://dx.doi.org/10.1016/j.fuel.2010.08.021]

[72] Dias MOS, Ensinas AV, Nebra SA, Maciel Filho R, Rossell CEV, Maciel MRW. Production of bioethanol and other bio-based materials from sugarcane bagasse: Integration to conventional bioethanol production process. Chem Eng Res Des 2009; 87(9): 1206-16.
[http://dx.doi.org/10.1016/j.cherd.2009.06.020]

[73] Vargas Betancur GJ, Pereira N Jr. Sugar cane bagasse as feedstock for second generation ethanol production. Part I: Diluted acid pretreatment optimization. Electron J Biotechnol 2010; 13(3): 10-1.
[http://dx.doi.org/10.2225/vol13-issue3-fulltext-3]

[74] Chandel AK, da Silva SS, Carvalho W, Singh OV. Sugarcane bagasse and leaves: foreseeable biomass of biofuel and bio-products. J Chem Technol Biotechnol 2012; 87(1): 11-20.

[http://dx.doi.org/10.1002/jctb.2742]

[75] Chandel AK, Antunes FAF, Anjos V, *et al.* Multi-scale structural and chemical analysis of sugarcane bagasse in the process of sequential acid–base pretreatment and ethanol production by Scheffersomyces shehatae and *Saccharomyces cerevisiae.* Biotechnol Biofuels 2014; 7(1): 63. [http://dx.doi.org/10.1186/1754-6834-7-63] [PMID: 24739736]

[76] Katakojwala R, Kumar AN, Chakraborty D, Mohan SV. Valorization of sugarcane waste: prospects of a biorefinery. Indust Municipal Sludge 2019; 47-60. [http://dx.doi.org/10.1016/B978-0-12-815907-1.00003-9]

[77] Bari MN, Ahmed T. Characterization of sugar industry wastes for solid state bioconversion. Cellulose 2011; 25: 38-3.

[78] Mohana S, Acharya BK, Madamwar D. Distillery spent wash: Treatment technologies and potential applications. J Hazard Mater 2009; 163(1): 12-25. [http://dx.doi.org/10.1016/j.jhazmat.2008.06.079] [PMID: 18675513]

[79] Karmee SK, Linardi D, Lee J, Lin CSK. Conversion of lipid from food waste to biodiesel. Waste Manag 2015; 41: 169-73. [http://dx.doi.org/10.1016/j.wasman.2015.03.025] [PMID: 25843356]

[80] Uçkun Kiran E, Trzcinski AP, Ng WJ, Liu Y. Bioconversion of food waste to energy: A review. Fuel 2014; 134: 389-99. [http://dx.doi.org/10.1016/j.fuel.2014.05.074]

[81] Wang ZJ, Lim BS, Wang ZJ, Lim BS. Electric power generation from treatment of food waste leachate using microbial fuel cell. Environ Eng Res 2017; 22(2): 157-61. [http://dx.doi.org/10.4491/eer.2016.061]

[82] Xin X, Ma Y, Liu Y. Electric energy production from food waste: Microbial fuel cells *versus* anaerobic digestion. Bioresour Technol 2018; 255: 281-7. [http://dx.doi.org/10.1016/j.biortech.2018.01.099] [PMID: 29428783]

[83] Matteson GC, Jenkins BM. Food and processing residues in California: Resource assessment and potential for power generation. 2005 ASAE Annual Meeting. 1. [http://dx.doi.org/10.13031/2013.19546]

SUBJECT INDEX

A

Abiotic 28, 29, 41, 42, 49, 52, 53, 54, 98
  responses 52
  stresses 28, 29, 41, 49, 52, 53, 54
Abnormal heart rhythms 95
ACC-deaminase 53
Acetobacter diazotrophicus 51
Acetogenesis 139
Acid 4, 5, 11, 26, 42, 49, 71, 82, 85, 110, 126, 147, 175, 191, 192, 193, 194, 200
  amino 26, 42, 49, 82, 85, 110, 147
  clofibric 175
  ferulic 200
  formic 126
  lactobionic 4
  neoaspergillic 71
  nucleic 110
  organics 192
  oxalic 194
  phosphoric 82
  succinic 200
  Sulphuric 5
  tartaric 193
Acidithiobacillus 11, 23, 26, 27
  ferrooxidans 11
Acinetobacter 146, 171
  baumanii 146
  radioresistens 171
Actinobacillus succinogenes 200
Actinomycetes 15, 44
Activated sludge 13, 14, 142, 175
  traditional 175
Activities 3, 7, 69, 70, 71, 93, 96, 97, 143, 165, 171, 200
  agricultural 97
  anti-bacterial 69, 70
  anti-fungal 71
  antioxidant 200
  chemolithotrophic 7
  cytotoxic 70
  ecological 3

hydrogenase 93
industrial 96, 165
phosphatase 171
reduced photosynthetic 143
Acute respiratory syndrome, severe 30
Acylated homoserine lactone (AHLs) 42, 44, 45, 47, 48
Acyl homoserine lactones 45
Adaptation progression 167
Adaptive responses, development of 53, 54
Adenosine tri-phosphate (ATP) 82, 93, 140
Adenoviruses 120
Adsorption 28, 106, 112, 143, 160, 178
  pollutant's 160
Aerobic 88, 139, 174
  bacteria-Thiosphaera pantotropha 88
  digestion 139
  process 174
Aeromonas 140
African trypanosomiasis 120
Agency for toxic substances and disease registry (ATSDR) 123
Agricultural 28, 91
  production 28, 91
  revolution 91
Agricultural lands 80, 84, 88, 89, 96, 152, 153, 175
  nutrient-deficient 96
Agricultural soils 80, 165
  productive 80
Agricultural waste 147, 188, 190, 196, 197
  developing 147
Agriculture 53, 84, 113, 137, 138, 188, 200
  industrial 84
  intensive inorganic 137, 138
  pollution 113
  sector 53
  wastes 188, 200
Agrobacterium tumefaciens 9, 45
Agrochemicals 152, 165
Agro-waste products 190

www.ingramcontent.com/pod-product-compliance
Lightning Source LLC
Chambersburg PA
CBHW050830220326
41598CB00006B/344